The Institute of Biology's
Studies in Biology no. 137

The Growth of Leaves

John E. Dale
B.Sc., Ph.D.
Reader in Botany
University of Edinburgh

Edward Arnold

First published 1982
by Edward Arnold (Publishers) Limited
41 Bedford Square, London WC1 3DQ

British Library Cataloguing in Publication Data

Dale, John E.
   The growth of leaves. – (The Institute of Biology's
   studies in biology, ISSN 0537–024– no. 137)
   1. Growth (Plants)   2. Leaves – Growth
   I. Title   II. Series
   581.3'1   ~~QK731~~

ISBN: 0 7131 2836 4

Photoset and printed by Photobooks (Bristol) Ltd

General Preface to the Series

Because it is no longer possible for one textbook to cover the whole field of biology while remaining sufficiently up to date, the Institute of Biology proposed this series so that teachers and students can learn about significant developments. The enthusiastic acceptance of 'Studies in Biology' shows that the books are providing authoritative views of biological topics.

The features of the series include the attention given to methods, the selected list of books for further reading and, wherever possible, suggestions for practical work.

Readers' comments will be welcomed by the Education Officer of the Institute.

1982
Institute of Biology
41 Queen's Gate
London SW7 5HU

Preface

Because of their role in intercepting and absorbing light, and converting it to chemical energy through the complex process of photosynthesis, leaves occupy a central position in plant growth studies, and hence in agriculture. This book summarizes our knowledge of the initiation of leaves and their early growth, and of lamina expansion; a brief treatment is also presented of the functioning leaf as it matures and develops towards senescence.

From Goethe onwards plant morphologists, anatomists, physiologists and biochemists have contributed to our understanding of leaf growth. But much remains to be discovered about the mechanisms involved in leaf morphogenesis and the reader will quickly realize that many intriguing questions will have to be answered before we can fully understand how leaves grow.

Edinburgh, 1982   J.E.D.

Contents

1 The Initiation of Primordia

1.1 The stem apex

No matter what their final size or shape all leaves have small beginnings. They originate as superficial bumps, *primordia*, produced in regular and predictable positions at the stem apex. For present purposes, the stem apex is defined to include the primordia of unexpanded leaves together with the associated stem tissues and the tip of the apex, the *apical dome*. The dome is often considered to approximate in form to a hemisphere or portion of a hemisphere, but it can vary widely between species in shape and size, neither of which are correlated with shape or size of the leaves initiated on it (Fig. 1–1). Also, since new primordia arise on the flanks of the apical dome, regular changes in its shape and size are inevitable during the course of leaf initiation.

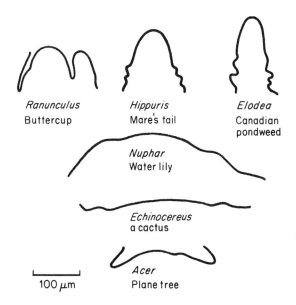

Ranunculus
Buttercup

Hippuris
Mare's tail

Elodea
Canadian
pondweed

Nuphar
Water lily

Echinocereus
a cactus

100 μm

Acer
Plane tree

Fig. 1–1 Outline drawings of apical meristems of six species to show variations in shape. (Redrawn from Wardlaw, C. W., 1957. *Amer. J. Bot.*, **44**, 176–85.)

Anatomically the dome usually shows a zonation between cells of the rather regular outer layers of *tunica*, and the less regularly arranged cells of the *corpus* (Fig. 1–2). The distinguishing feature of the tunica is that in it the plane of cell

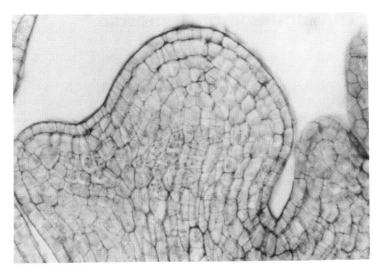

Fig. 1-2 Section through the stem apex of *Pisum sativum* to show primordial initiation. The outer tunica is especially clear. (Photograph and section by Dr R. F. Lyndon.)

division is almost exclusively anticlinal, i.e. normal to the plane of the tissue, whereas in the corpus planes of division are apparently random. Too much significance should not be attached to the number of layers in the tunica as this can vary during the interval between appearance of successive primordia, the *plastochron*. Because of this it is unrealistic to argue for the involvement of a fixed number of tunica layers in primordium initiation. What is important to note is that primordium initiation involves the superficial layers of the apex and does not involve deep-seated tissue as is the case in lateral root initiation.

Measurements of growth of the stem apex are often difficult and invariably laborious to obtain. Because of their small size and irregular shapes, dimensions of the dome and primordia are usually best obtained using sectioned material examined under the microscope (WILLIAMS, 1975). From the study of serial sections it has been shown that there is a gradient in growth measured in terms of size, with low rates at the summit of the dome and higher rates on the flanks. These differences are seen also in cell division rates which tend to be two to three times higher at the base of the dome than at the summit. In other experiments small carbon particles have been applied to the tip of the dome. As growth occurs, the particles are dispersed downwards and measurements of the displacement rates at different positions on the flanks have confirmed a faster growth rate towards the base of the dome.

Although growth rates are not uniform throughout the dome this variation is not in itself the factor controlling dome shape which is considered to result from the balance between growth in the longitudinal and latitudinal directions (LYNDON, 1976).

Immediately following initiation of a new primordium, the volume of the apical dome is minimal. It then increases to reach a maximum immediately prior to initiation of the next primordium whereupon the cycle is repeated. Now suppose that there is a change in dome size between successive plastochrons such that the size at the beginning of the second plastochron is larger than that at the beginning of the first. Such a change can be brought about in a number of ways depending upon plastochron length, dome growth rate and size of the primordium at initiation. For example, if dome growth rate and size of the primordia on initiation are constant but the plastochron is lengthened then size of the dome between plastochrons will increase. Again, if the plastochron and dome growth rate are constant but size of the initiated primordia goes down, then size of the dome between plastochrons will increase. On the other hand, if dome growth rate and volume are constant but primordium size goes up then the plastochron will shorten. Growth rate, dome and primordium sizes and the plastochron are interrelated and a change in any one will affect the measured value of at least one of the other parameters.

1.2 Hypotheses of primordial origin

The fact that a gradient of increasing growth rate exists from summit to base of the apical dome might be thought significant since primordial inception could be attributed to the higher rate of growth. However, to decide this requires detailed information which is available for very few species. In pea and tomato those positions on the dome occupied by cells which will form the (incipient) next and next-but-one primordia (designated I_1 and I_2) each show a distal region with slow cell division rates and a basal region with a faster rate, and it could be argued that the primordium arises as a result of faster growth in the basal region. However, for the pea it appears that primordial origin is brought about primarily by changes in the direction of growth rather than by differences in growth rate. Study of apices at all stages of the plastochron has shown that only anticlinal divisions occur in the apical dome during the first two-thirds of the plastochron. Subsequently periclinal divisions, parallel to the plane of the surface, are seen and at I_1 anticlinal divisions become restricted to the outer tunica layer. That is to say, at the time when initiation of I_1 is about to occur there is a marked change in the direction of growth suggesting that some constraint, allowing only anticlinal division, is lifted (LYNDON, 1976). It is the onset of random growth with division in all planes that enables the primordial buttress to develop as a bulge beyond the margin of the apical dome.

If it is the direction of growth that is important in the initiation of a primordium, is there any significance in the faster growth rate of the basal portion of I_1 and I_2? It is a feature of primordial growth that the abaxial surface, further from the dome, grows faster than the adaxial side. The result is that the primordium tends to curve upwards. It is attractive to think that this difference in growth of the two sides is established by the initial differences in growth rate in the incipient primordium.

One of the earliest hypotheses of primordial origin, going back to the nineteenth century, was that growth rates on the outside of the dome are higher than those in the underlying tissues and that as a result of the pressures arising from this the superficial layers are wrinkled, leading to the formation of a primordial initial. This idea, that the outer layers are subject to pressure has been tested by making shallow incisions on the dome at the site of I_1 and I_2. In many cases where this has been done the incisions, far from remaining closed due to forces of compression, have tended to gape open. Thus the idea of surface compression is not supported. However, a basic problem with surgical treatments of this kind is to distinguish the effects caused by wounding from the phenomenon being examined.

Another theory of primordial origin with a long history is the field or repulsion theory. The great German botanist Hofmeister, more than a hundred years ago, argued that the position at which a primordium was found was fixed by the location of the two or three previously formed primordia. This idea has been developed by a number of workers from Schoute in 1913 to THORNLEY (1976) and as postulated involves inhibition of primordial formation. A summary of the modern view is as follows:

(1) The stem apex contains an inhibitory substance, or morphogen, which at suitably high concentrations prevents formation of new primordia.

(2) The sites of production of this inhibitor are the older primordia and also the cells at the summit of the apical dome.

(3) The inhibitor diffuses isotropically (i.e., equally in all directions) from sites of synthesis, and may also be subject to breakdown in the tissues of the apex.

(4) At points where the inhibitor concentration falls below a threshold level a new primordium will arise; inevitably such points will be distant from the apex summit and from the older primordia.

Despite the fact that no inhibitory morphogen has yet been isolated the field

Fig. 1–3   The production of plantlets by *Bryophyllum tubiflorum*. Three or four plantlets are normally produced at each leaf tip.

theory has received considerable support, the most recent coming from
modelling studies involving computer simulation. In one such analysis leaf
arrangement in *Bryophyllum tubiflorum* was modelled. In this species adventitious plantlets are formed at the tip of leaves (Fig. 1–3), and the first leaves found
on the plantlets appear to be paired and opposite; subsequently leaf arrangement undergoes a series of changes culminating in either tricussate or spiral
arrangements in larger plants. Anatomical studies have shown that the first
leaves are neither exactly the same size, nor inserted exactly opposite each other.
This initial assymetry is important for it implies that the concentration of
inhibitory morphogen in the primordia will differ. A computer simulation in
which values were assigned to apex width and growth rate, and to diffusion and
decay coefficients of the supposed inhibitor, together with estimated threshold
concentrations necessary to allow leaf initiation, gave good approximations to
the pattern of leaf arrangement actually found. Furthermore, the model yielded
this pattern when programmed for an increase in apex size, a condition which
occurs *in vivo*. Therefore there is no need to argue that changes in leaf
arrangement necessarily result from changes in rate of production or breakdown
of the inhibitor; changes in apical size alone will suffice to change primordial
initiation pattern provided there is an initial assymetry in the arrangement of the
first primordia. The increase in dome size which occurs as a plant ages may thus
be of significance to leaf initiation as well as to leaf shape and the transition to
flowering.

1.3   Phyllotaxis

In a large number of species leaves are arranged singly or in pairs on a spiral
about the stem axis (Fig. 1–4). This regular arrangement or *phyllotaxis* can be

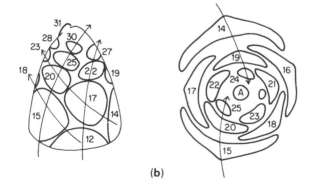

(a)   **(b)**

Fig. 1–4   The arrangement of leaf primordia on the stem of flax seen (a) in face view, (b) in
transverse section across the stem apex. The solid lines are the parastichies and the (3+5)
phyllotaxis is clearly seen in (a). (Drawings based on Williams, R. F., 1975. *The Shoot Apex
and Leaf Growth*. Cambridge University Press.)

accounted for by primordium-initiating mechanisms of the type postulated by the field theory. Early descriptions of phyllotaxis made much of the *divergence angle* formed between centres of primordium P_n, the apical dome, and primordium P_{n+i}. This is close to the Fibonacci angle of 137.5°. From examination of apices (Fig. 1–4) it can be seen that primordia fall in sets of spiral patterns or *parastichies*. For the example shown two main sets of spirals can be seen running in opposite directions round the apex. In the first set successive primordia differ in serial number by 3 (e.g. 14, 17, 20. . .), while in the other the difference is 5 (e.g. 15, 20, 25 . . .). Such a phyllotactic arrangement is often described as (3 + 5). Other arrangements are known, many of which can be described by two successive terms in the Fibonacci series (1, 1, 2, 3, 5, 8, 13, 21, 34, 55 . . .).

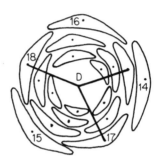

Fig. 1–5 Transverse projection of the stem apex and primordia of flax. The plastochron ratio can be derived by measuring the distance from the centre of the dome, D, to the centres of successive primordia. (Redrawn from Williams, R. R.F., 1975. *The Shoot Apex and Leaf Growth*. Cambridge University Press.)

Descriptions of phyllotaxis based on divergence angle take no account of growth of the apex and of the vertical displacement of primordia that occurs during development. RICHARDS (1951) pointed out that by joining the centre of successive primordia an exponential spiral, the so-called *genetic spiral*, is generated. Using transverse projections (e.g. Fig. 1–5) the distances from the centre of the apex, D, to centres of the primordia can be measured and the *plastochron ratio*, r, obtained:

$$r = \frac{x_n}{x_n + 1}$$

here $\dot{x}_{n_1}$ and x_n are the distances from apex centre to centres of primordia of serial number $x + 1$ and x respectively. The plastochron ratio measures the relative distance by which primordium x grows from the centre of the apex in one plastochron. The *relative rate* (see HUNT, 1978) at which primordia recede is given by $\log_e r$ which is the relative radial growth rate per plastochron. On the assumption that relative growth rate in the vertical plane is similar to the radial value, relative growth rate in volume of the apex will be given as 3. $\log_e r$ per plastochron.

The plastochron ratio is thus a useful tool for growth analysis of the stem apex. Together with the divergence angle, r defines the transverse component of the apex/primordium system. RICHARDS (1951) describes in detail how by using in addition a measure of the angle of the cone tangential to the apex a full quantitative description of phyllotaxis can be given.

Attempts to change phyllotaxis by application of growth regulating substances have usually failed but SCHWABE (1971) was able to induce an alteration in phyllotaxis from spiral to paired and opposite by application of tri-iodobenzoic acid to *Chrysanthemum* stems. This treatment had only slight effects on relative growth rate of the whole apex but did cause successive primordia to be situated lower on the flanks of the apex, suggesting an increase in the length of the presumptive internodal tissue. On the field theory such an increase might be expected to change inhibitor distribution and concentration and thus to affect divergence angle and phyllotaxis.

In many plants the phyllotactic arrangement is not constant throughout the life cycle. Thus the paired opposite arrangement of the cotyledons may be followed by changes to give a spiral, helical arrangement of progressively higher order, resulting ultimately in the characteristic arrangement of the floral structures. The causes of these changes are not known with any certainty. They

Fig. 1–6   Leaf display in *Poinsettia*. The view is from above and the effects of phyllotaxis and variable petiole length on leaf arrangement are clearly seen. (Photograph by W. Foster.)

can be explained on the field theory in terms of assymetry in the apical system such as was found for *Bryophyllum tubiflorum*. An alternative possibility is that leaf arrangement may be controlled by arrangement of the pro-vascular tissues entering the apical bud. This is an attractive idea, for the apex is completely dependent upon the supply of metabolites from below it and these are transported along the vascular path; consequently local variations in supply may well be dependent upon the precise location of the vascular system. Unfortunately, there is an unresolvable chicken-and-egg dilemma here for while it may be that the pro-vascular path determines the site of the primordium, it is at least as likely that the primordium determines the pattern of differentiation of the vascular tissues.

The initiation of leaf primordia is an ordered process, the regulation of which leads to the phyllotaxis characteristic of the species. A major and important consequence of phyllotaxis is that overlap between adjacent leaves is minimized; coupled with petiolar movements which can alter leaf orientation, this means that leaf display is optimized for the interception of light (Fig. 1-6).

1.4   The fate of primordia

The root produces lateral organs of one type which are similar to itself. This is not the case for the stem which initiates lateral organs of two types, of limited and unlimited growth. Those in the first category include as well as foliage leaves, scales, bracts and floral structures; those in the second are lateral stems which are initiated as axillary buds. The mechanisms involved in the initiation of these different structures have been investigated.

Because of their comparatively large size apices of ferns such as *Dryopteris* (Fig. 1-7a) have been a favourite choice for experiments on primordial determination. However the pteridophyte apex differs from that of the angiosperm since it grows from a large apical cell rather than from the dome-shaped meristem characteristic of the latter. In a long series of studies using surgical treatments on fern apices, C. W. Wardlaw was able to show that isolation of the presumptive leaf primordium from the apical cell, and from adjacent leaf primordia (Fig. 1-7b), led to its development as a bud. Other experiments showed that young initiated primordia could also be induced to form buds when isolated from the apex by similar treatments. These incipient and newly-initiated primordia are thus undetermined in that while they normally give rise to leaves, they nevertheless have the potential to develop as buds. A number of attempts to demonstrate this phenomenon in angiosperms have failed and here it would appear that primordial fate is determined by the time of initiation, or perhaps even earlier.

The results of the experiments with fern primordia have led to the suggestion that in normal development, chemical factors originating in the apex direct growth along a route to form a leaf. This idea has been investigated by using the cinnamon fern, *Osmunda cinnamomea*. When primordia of this fern are excised from the apex and grown on agar medium their development depends upon age at excision and on the proximity to other primordia. When very young

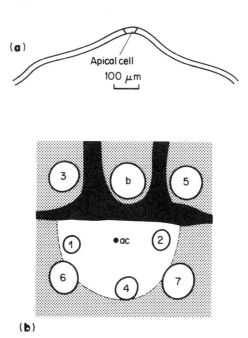

Fig. 1-7 The apex of the fern *Dryopteris* in outline (**a**) and the pattern of cuts used to isolate the incipient primordium which developed as a bud (**b**); previously formed primordia are numbered 1-7. (From Wardlaw, C. W., 1949, *Nature*, **161**, 167-70.)

primordia are isolated they will most often develop as shoots unless older primordia are also present in which case development as leaves is most common. When rather older primordia are cultured they often form a shoot bud at the base while remaining recognizable as leaf primordia. The oldest primordia invariably give rise to leaves when cultured.

Work of this kind involving isolation and culture of primordia is technically difficult and unfortunately the results obtained are often difficult to interpret. Nevertheless the possibility of chemical morphogens coming from the apex and directing primordial development is supported by these studies on ferns.

While the initiation of floral structures at the apex is too large a subject to be considered here it should be noted that this is seldom an abrupt process since gradual changes in leaf size, and often shape, usually precede the appearance of flowers. Thus the mechanisms responsible for flower initiation may also affect primordial fate and leaf development, perhaps secondarily.

A transition in shape, size and photosynthetic activity of foliage leaves is also often the prelude to production of scale leaves, or *cataphylls*. Such structures are usually initiated late in the growing season in response to daylength and temperature changes and serve to protect the bud over winter. Their initiation is similar to that of foliage leaves but subsequent development of the primordium is brief and followed by rapid maturation. There are reports that applications of abscisic acid can induce bud scale formation and it is likely that accumulation of abscisic acid under the shortening days of late summer, and its movement to the apex, is associated with switching primordial development to the direction of the scale leaf.

1.5 Seasonal variation in primordial production

For annual plants the production of foliar primordia at the stem apex is continuous until the advent of unfavourable environmental conditions, usually low temperatures, or until the apex begins to produce floral structures. In both cases death of the plant eventually follows. For perennial plants the stem apex usually functions over at least two growing seasons, often over many more. For the herbaceous perennial, axillary buds are initiated during the first season, and these overwinter before elongating in the second season when pre-formed primordia will unfold, and perhaps new primordia develop to give leaves; at the end of the second season the apex will die and new lateral shoots will develop to continue growth in subsequent seasons. In woody perennials, the apical meristem may be long-lived and produce primordia over many years. The temporal pattern of primordial production varies greatly in such species. In some, like the apple (Fig. 1–8) primordia are produced during the first season and a resting bud is formed, the bud scales of which are formed from the early-initiated primordia. During the second season, the foliar primordia develop further and the leaves unfold; depending on its position on the stem the apical meristem may repeat the process, or alternatively may initiate floral primordia. In other species, such as the oak, several flushes of growth are seen with leaves

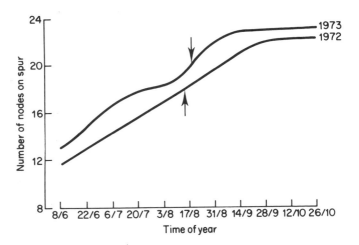

Fig. 1–8 The production of primordia on shoots of apple in two successive seasons. The arrows indicate the time when flower primordia began to be initiated. (Redrawn from Abbott, D. L., 1976. *Ann. Rept. Long Ashton Res. Stn.*, 171.)

unfolding in April, June, and even in August; these leaves are developed from primordia initiated during or just after the previous flush. Some species, such as *Eucalyptus*, show a continual production of primordia without the formation of resting buds, but with the plastochron lengthening during adverse environmental conditions.

Accepting the need for the deciduous habit in broad-leafed trees of the cool-temperate regions, the winter bud can be seen as a device enabling primordia to be initiated and checked in development until such time as favourable conditions allow rapid near-synchronous lamina expansion to occur.

2 Early Growth

2.1 Meristematic activity in the young primordium

That the development of the leaf is an orderly and regulated process is clearly seen in early growth of the primordium. Three main phases can be identified:
  (1) establishment of the foliar buttress
  (2) establishment of the primordial axis
  (3) initiation of the lamina.

The newly-initiated primordium appears as a bulge on the flank of the apex, and often soon extends some way laterally around it. During its earliest development the primordium is awl-shaped, oriented horizontally and appressed to the apex; at this stage, when the ratio of vertical height to horizontal width is minimal, the primordium is often referred to as the *foliar buttress*. Lateral enlargement of the foliar buttress continues over a long period but the major direction of growth gradually changes to the vertical with the extension of a peg-like pro-axis from the upper surface of the buttress. In the case of many compound leaves, establishment of the axis of the main leaflet is closely followed by development of centres of growth from which lateral leaflets arise (Fig. 2–1).

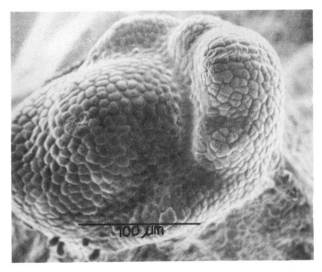

Fig. 2–1   A scanning electron micrograph of the stem apex of *Phaseolus vulgaris* showing the initiation of the primordium of the first trifoliate leaf 48 h after planting. Note the presence of stipule primordia and of the initials of the lateral leaflets. (Photograph by Dr Fatima Aleixo-Pereira.)

Organization of the extending primordium in meristematic terms is quite complex. Growth of the primordial axis is usually most rapid at the tip suggesting a discrete meristem there. However, meristematic activity also occurs along the length of the primordium and gradually this intercalary growth comes to predominate and apical growth declines and eventually ceases.

Vascular tissues are developed from a procambium (§2.4) and there is often a more-or-less clearly defined adaxial meristem whose activity leads to lateral extension of the primordium and increase in width. In addition, the meristems associated with lamina initiation develop while the primordium is still small.

2.2 Lamina initiation

Usually by the time that the primordium is between 0.1 mm and 1 mm long, lamina initiation commences with the appearance of meristemic regions at the lateral flanks of the primordium. In the case of sessile, entire leaves these *marginal meristems* extend almost to the base of the primordium, but for petiolate leaves they extend only part of the way along the axis. In some deeply

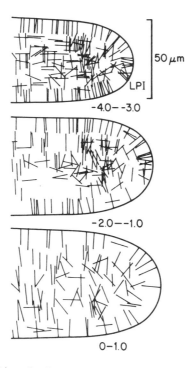

Fig. 2–2 Planes of division of cells at the margin of developing leaves of *Xanthium strumarium* at three stages of development indicated by values of LPI. (From Maksymowych, R., 1973. *Analysis of Leaf Development*. Cambridge University Press.)

indented and compound leaves the marginal meristem is not continuous along the primordial length but is interrupted at intervals, and as a consequence lamina proliferation is discontinuous.

The marginal meristem consists of two classes of cells distinguished by position and plane of division. The outermost *marginal initials* form a single layer of cells which divide almost entirely in the anticlinal plane and extend the epidermis of the blade. Immediately below the marginal initials is a layer of *submarginal initials* whose divisions are in the periclinal or oblique planes (Fig. 2–2); these initials contribute cells to the inner layers of the leaf.

Just as apical growth of the primordium plays an important but temporary role in primordial extension, so activity of the marginal meristems may be comparatively short-lived. It is complemented and superseded by generalized meristematic activity throughout the developing blade which becomes a *plate meristem*. As a consequence of this the majority of cells in a mature leaf arise from divisions at sites distant from the leaf margin.

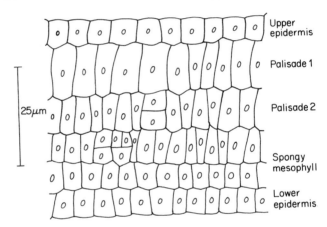

Fig. 2–3   The layered arrangement of cells in an expanding primary leaf of *Phaseolus vulgaris*.

The developing leaf has a characteristic layered structure (Fig. 2–3) the origin of which lies in the pattern of cell division at the leaf margin. The predominantly anticlinal plane of division of the marginal initials has already been mentioned. The periclinal divisions in the submarginal initials lead to an increased number of cell layers but there is no evidence for the earlier belief that cell lineages can be derived from identifiable initials in the submarginal region. Nor is the pattern of periclinal divisions rigidly controlled except inasfar as a more-or-less regular number of cell layers is ultimately formed. Periclinal divisions are not unique to the marginal meristem but are found also throughout the plate meristem although here the predominant plane of division is anticlinal.

2.3   Primordial development in monocotyledons

The description given above applies to the 'typical' dicotyledon leaf. The best known type of leaf development in the monocotyledons is that of the grass leaf. Since a full account is given by LANGER (1979) the description here is brief.

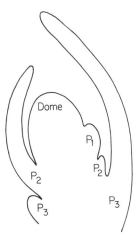

Fig. 2–4   Outline drawing of the shoot apex of barley (*Hordeum vulgare*) showing the three youngest foliar primordia. Note how lateral extension of the primordial buttress results in its appearance on both sides of the apex.

The grass leaf begins development as a single ridge, oriented laterally on the apical flank (Fig. 2–4). Early growth of the primordium resembles that already described except that considerable lateral extension of the foliar buttress occurs to give a girdle of tissue around the stem apex, open on the distal side opposite the point of origin of the primordium. The primordium is thus crescent – shaped extending as a result of growth at the distal edges analogous to marginal growth in the dicotyledon leaf, and eventually becomes a hood-shaped structure overtopping the apical dome and younger primordia. At about this time intercalary growth becomes established in a discrete meristem at the base of what will become the blade. A second intercalary meristem arises below the first and gives rise to the leaf sheath. In contrast to the blade meristem which cuts off cells towards the leaf tip, or acropetally, the sheath meristem cuts off cells basipetally and it commences and ends its activity rather later. Meristematic activity is thus highly localized and there is no plate meristem as found in the dicotyledons.

But a range of different types of leaf development is found in the monocotyledons. Laminate leaves are found in a number of families including Gramineae, Amaryllidaceae and Liliaceae, but in others, such as Juncaceae and Iridaceae, *unifacial* leaves are found. These are leaves which are either tubular as in *Juncus* or *Allium*, or flattened in the median plane as in *Iris*. Characteristically, the vascular bundles in unifacial leaves form a ring or are arranged in two parallel rows. This variation in leaf structure has excited considerable and sometimes acrimonious controversy among plant morphologists. A recent interpretation by KAPLAN (1973) suggests that whereas in the dicotyledons the distal tip of the primordium gives rise to the lamina while the middle and proximal regions give rise to petiole, and leaf base and stipules respectively, in the monocotyledons the tip of the primordium gives rise to the unifacial part of the leaf where present, while lamina, petiole and leaf sheath are all derived from the proximal portion of the primordium. In laminate leaves the primordium tip remains rudimentary and can sometimes be identified by a short spine as in some species of *Sanseveria*, mother-in-law's tongue.

2.4 Vascular development

There is a close relationship between the arrangement of vascular bundles in a stem and the pattern of primordial initiation. Vascular strands are also important as channels for the supply of water and metabolites to the growing leaf, and for the export of photosynthetic assimilates out of the mature leaf.

Although species vary in detail the main features of vascular development in the young leaf are similar. When primordial elongation begins, or very soon after, a procambial strand, continuous with that of the stem is closely associated with the foliar buttress (Fig. 2–5); this strand develops acropetally to reach the

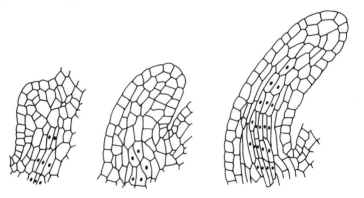

Fig. 2–5 The development of the procambial strand during the early growth of a leaf primordium of flax. Dotted cells are procambial elements. (Redrawn from Girolami, G., 1954, *Amer. J. Bot.*, **41**, 264–73.)

tip of the primordium by the time that apical growth has ceased. The various elements in the strand also develop from the base, with phloem differentiating in advance of xylem, both temporally and spatially. The mid-vein continues to enlarge both in length and in radius as the leaf enlarges, with the continual production and differentiation of new procambial elements. Second and third order veins develop outward from the base of the primordium and may also mature acropetally. However, the highest order veins and veinlets tend to mature earliest in the distal region of the blade. There is structural continuity between the main veins since the procambium of a vein and that of its subtending higher-order vein is continuous; this is not always the case for the minor veins and structural discontinuity during development may occasionally lead to veinlets not being connected to the main network. Vein development is complete, except for a few of the minor veins at the base of the leaf, by the time the blade is about 60–80% of its final size.

An interesting observation is that in species which have internal phloem in their vascular bundles (e.g. the potato and cucumber families, Solanaceae and Cucurbitaceae) this phloem develops late in blade expansion. There is some experimental evidence that internal phloem is associated with the export of photosynthetic assimilate out of the leaf. Other evidence indicates that loading of assimilate into phloem of the leaf occurs at the endings of the minor veins which are the last parts of the vascular system to develop. The relationships between leaf development and export of assimilates is discussed in Chapter 6.

2.5 Primordial growth rates

Because of their size it is not easy to measure the growth of primordia without using destructive methods of which the most satisfactory is that of serial reconstruction. This involves cutting thin serial sections through apices of known ages and painstakingly reconstructing the shape, and size, of the primordia from these. It is an intensely laborious method but has yielded valuable data on primordial volumes and growth rates in a number of species.

Two main patterns of growth of primordia have been found (Fig. 2–6), both involving a period when primordial volume increases logarithmically with time. In species such as tobacco and clover, this exponential growth of the

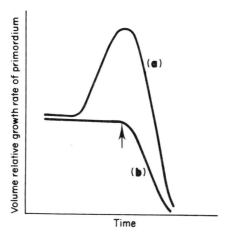

Fig. 2–6   The time course of relative growth rate for primordia of (a) wheat and (b) tobacco. The arrows indicate onset of leaf emergence and lamina unfolding.

primordium continues until around the time of leaf emergence and lamina unfolding when the volume relative growth rate begins to fall sharply. The second type of pattern is shown by wheat, and here the exponential phase is followed by a short period of much higher growth rate, which again declines as leaf emergence and unfolding proceeds.

During its development a primordium is in contact with others, both younger and smaller, and older and larger, and it may be subject to physical pressure and constraint from these primordia. WILLIAMS (1975) has suggested that the increase in volume relative growth rate for wheat primordia at the end of the exponential phase results from the primordium enlarging to a size where it suddenly escapes from the physical restrictions imposed by the older leaves, perhaps by beginning to emerge above the sheath of the next older leaf. This interesting idea highlights the fact that the close physical association between primordia at the apex may well affect both the pattern and rate of their growth. But it does not explain why growth of primordia is exponential over a substantial period, nor why the exponent is often found to decrease for successive primordia.

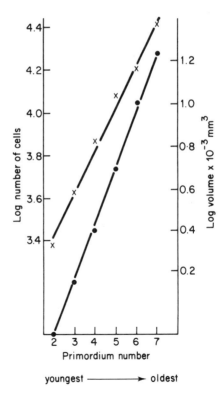

Fig. 2–7 Cell numbers (x) and volumes (●) in successive primordia at the apex of *Lupinus albus*. (From Sunderland, N. and Brown, R., 1956, *J. exp. Bot.*, 7, 127–45.)

Primordial growth rate is reflected closely in changes in cell number (Fig. 2–7) which are a consequence of the intense meristematic activity already discussed (§2.1). Cell number increase is often apparently exponential and could result if all or most of the cells in the primordium continue to divide, as seems to be the case. Increases in mean cell volume make a much smaller contribution to growth of the primordium than changes in cell number. The increase is mainly due to enlargement of comparatively few cells, mostly in the procambial tissue, rather than to a generalized increase in size of all cells. At the end of the phase of primordial growth, when the leaf is about to unfold, average cell size remains much less than that of cells in the mature leaf; cell number too, is very small compared with the final value achieved.

There is limited evidence that as the apical dome enlarges with age, the size of successive primordia at initiation may also increase. Clearly, to maintain the same relative growth rate as for smaller primordia a greater supply of metabolites is required and if the amount is restricted growth rate must be reduced. Such an interpretation could explain the lower growth rates of successive primordia, but direct evidence to support it is difficult to obtain.

3 Expansion of the Lamina

3.1 The plastochron index

Even when grown from apparently uniform seed, and under carefully controlled conditions, plants of the same age may vary considerably in their morphology and physiology. To overcome these differences, and to express age on a developmental basis rather than on a chronological one, ERICKSON and MICHELINI (1957) proposed a *plastochron index* based on simple, non-destructive measures of leaf length. This index of development depends upon three assumptions (see Fig. 3–1):

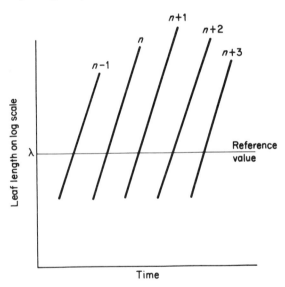

Fig. 3–1 Idealized growth patterns for five successive leaves to illustrate assumptions necessary for the calculation of the plastochron index.

(1) that the time interval between appearance of successive leaves on the same plant is constant

(2) that early growth in length of a leaf occurs at an exponential rate so that the plot of the logarithm of leaf length against time gives a straight line

(3) that early growth of leaves on the same plant shows similar relative rates.

For species where these assumptions hold the plastochron index is calculated as

$$PI = n + \frac{\log_e L_n - \log_e \lambda}{\log_e L_n - \log_e L_{n+1}}$$

where n is the serial number on the stem of the youngest leaf whose length exceeds that of the reference value λ (often taken as 10 mm), and L_n and L_{n+1} are lengths, in mm, of leaves n and $n+1$. As an example of the calculation consider a plant whose 5th and 6th leaves up the stem have lengths 12 and 7 mm respectively; let the reference length, λ, be 10 mm, then

$$PI = 5 + \frac{\log_e 12 - \log_e 10}{\log_e 12 - \log_e 7} = 5.338$$

The plastochron index is a measure of the developmental age *of the plant*. It is possible from this to obtain the developmental age of any *leaf, i*, by calculating the leaf plastochron index, LPI, as

$$LPI_i = PI - i$$

where i is the serial number of the ith leaf. Thus in the example above LPI for leaf 4 will be 1.338, for leaf 5, 0.338, and for leaf 6, −0.662. A negative value of LPI indicates a leaf of length less than the reference value, and a positive value a length greater than it.

The use of the word 'plastochron' in these indices is somewhat unfortunate since it will be clear that they are based not on the interval between initiation of successive leaves, but on the interval between attainment of the reference length. Nevertheless both plastochron index and leaf plastochron index have been widely used in developmental studies on leaves. In the cocklebur (*Xanthium strumarium*) the use of LPI has added greatly to the description of leaf growth in cellular terms, and in the cottonwood (*Populus deltoides*) the index has been used as a basis for elegant studies on vascularization, photosynthesis and translocation in developing leaves. However, in some studies the index cannot be used, either because of difficulties in getting an appropriate measure, as in grass leaves where because of the mode of extension of the leaf a reference length cannot be measured, or when the basic assumptions do not hold.

Table 1 The relationship between LPI and leaf size for the 9th leaf of cocklebur, *Xanthium strumarium*. (From Maksymowych, R. and Erickson, R. O. (1960). *Amer. J. Bot.*, **47**, 451-9.)

Stage of development	Leaf size	LPI
Lamina initiation	220 μm long	−4.8
Leaf unfolding commences	5–10 mm long	−0.5 to 0
Leaf expansion completed	125–150 cm² area	6 to 7

3.2   Increase in lamina size

The importance of the phase of lamina expansion to leaf growth is shown by data in Table 1 for the leaf of *Xanthium*, which shows a 250-fold increase in area following emergence from the bud. Growth in length and area is often sigmoid when plotted against either time or LPI, but for some species the initial exponential phase may be either absent or very short-lived. In the case of the paired primary leaves of the french bean, *Phaseolus vulgaris* (Fig. 3–2) lamina unfolding begins on day 6 or 7 but relative growth rate for area expansion shows a continual decline with time, which is very close to a negative exponential. The significance of this fall is discussed below in terms of changes in number and size of cells.

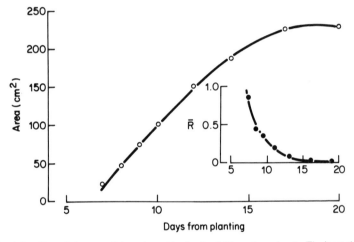

Fig. 3–2   Growth in area of the primary leaf pair of *Phaseolus vulgaris*. The inset shows the decline in relative growth rate with time.

Lamina expansion is not uniform over the whole of the developing blade and some regions develop faster than others, as was shown many years ago by AVERY (1933) who marked expanding tobacco leaves with Indian ink squares. Avery concluded that there was a gradient of maturation along the leaf from tip to base, so that the basal regions grew fastest and longest. Growth also tends to be faster in central parts of the leaf close to the main vein. More recent data showing these trends for growth in area are seen in Fig. 3–3, for leaves of *Xanthium* at three developmental stages. In the grass leaf, and in narrow leaves of dicotyledonous species such as *Plantago* (the plantain) the basal regions of the leaf mature considerably later than the tip establishing a very marked physiological and developmental gradient along the blade. Another point of interest is that increase in lamina thickness often continues after expansion in area has ceased. Usually, this is the consequence of continued extension of cells of the palisade mesophyll and tends to be most marked in conditions of high light intensity (see §5.3).

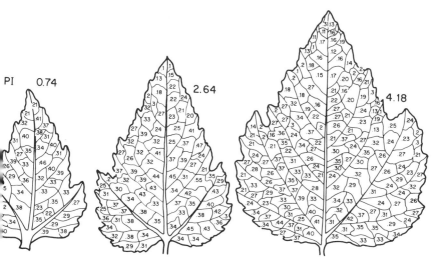

Fig. 3–3 Relative growth rates for expansion of different regions of a leaf of *Xanthium strumarium* at three developmental stages. (From Maksymowych, R., 1973, *Analysis of Leaf Development*. Cambridge University Press.)

3.3 Changes in cell number

For a long time it was thought that laminar expansion was a process whereby cells previously formed in the primordium merely enlarged as the leaf unfolded. With the development of accurate methods for estimating cell numbers, e.g. by macerating tissues and counting the separated cells, this view has been shown to be wrong. The majority of cells in a leaf are formed after expansion of the lamina has begun. This is shown very clearly in data for leaves of sunflower and *Xanthium* (Fig. 3–4). For the sunflower, leaf cell number at unfolding is about 2.5 million but when the leaf is fully expanded cell number is about ten times this value. In the case of *Xanthium*, the change in cell number during expansion is even more spectacular for between LPI 0 and LPI 2.5 there is a 100-fold increase. Table 2 gives estimates of numbers of cells formed after leaf unfolding has commenced in a number of species.

The massive cell number increase which occurs during leaf expansion results from meristematic activity throughout the whole of the plate meristem (§2.2). But, as for blade expansion in general, there is evidence for a basipetal gradient in cell division with divisions ceasing first at the tip of the leaf and last at the base. The extreme form of this trend is seen in the grass leaf where divisions occur only in the basal meristem at the proximal end of the blade so that extension of the lamina is by cell enlargement alone.

As well as this commonly-observed basipetal gradient in cell division along the leaf, there are also differences in the duration of division in the various layers of the leaf. With the exception of cells giving rise to stomata, division ceases first

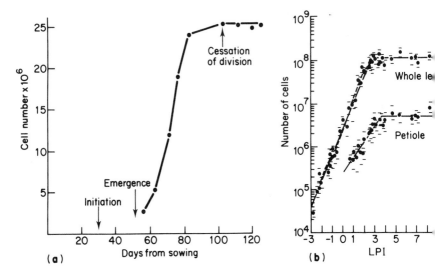

Fig. 3-4 Changes in cell number during development of (a) the 10th leaf of sunflower, *Helianthus annuus* and (b) the 9th leaf of *Xanthium strumarium*. ((a) From Sunderland, N., 1960, *J. exp. Bot.*, **11**, 68–80; (b) from Maksymowych, R., 1973, *Analysis of Leaf Development*. Cambridge University Press.)

in cells of the epidermis and continues longest in those of the palisade mesophyll; divisions of cells in the stomatal complex can continue for some time after other cells in the epidermis have ceased to divide.

Thus cell division is not uniform over all parts of the expanding leaf and there are variations in the duration of the cell division phase. An underlying question of considerable interest concerns the mechanisms responsible for causing divisions to cease in a particular part or layer of the blade. Although it is an unanswered question, the trends in cell number increase are informative. The data suggest that increase in cell number in the primordium is exponential. Such

Table 2 The percentage of cells in a mature, fully-expanded leaf that are formed during lamina expansion.

Plant	Percentage
Cocklebur (*Xanthium*)	99
Tobacco (*Nicotiana*)	99
Sunflower (*Helianthus*)	90
Clover (*Trifolium*)	60
Cucumber (*Cucumis*)	70–98
French bean (*Phaseolus*)	
Primary Leaf	17–35
Trifoliate Leaf	75–90

a relationship could result if a constant proportion of cells present go into successive divisions, with cell cycle time (the time from one prophase to the next) being constant; it could equally well be obtained if a variable proportion of cells entered successive divisions, but with compensatory changes in cell cycle time to maintain the linear relationship. In the case of *Xanthium*, cell division in the primordial and early expanding phases is exponential (Fig. 3–4); but exponential increase in cell number during expansion is not always found and in many species the rate of increase in cell number falls steadily to zero in expanding leaves. This decline could be explained if a smaller and smaller proportion of cells enter each succeeding division, with constant, or even lengthening cell cycle time. There is indirect support for this view and if it is correct, one has to ask how it is brought about. Some evidence suggests that cell number in a leaf is dependent upon conditions experienced by older leaves (DALE, 1976), and that developing leaves may depend upon import from older leaves of factors necessary for cell division. If this is so, competition for such factors might occur between different parts of the leaf, with those parts closest to the import channel, the vascular tissues, being most favoured. On this hypothesis, cells at the periphery of the leaf, at the tip, and in the epidermis, would be at a disadvantage and would be the first to cease dividing. This interpretation is speculative and the existence, let alone the identity, of the metabolites required for the continuation of cell division remains to be demonstrated.

3.4 Cell enlargement

During lamina expansion very high rates of cell enlargement occur and the increase in mean cell volume over this phase can vary between 10- and 35-fold according to species and cell type. MAKSYMOWYCH (1973) has conducted extensive

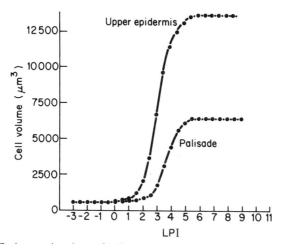

Fig. 3–5   The increase in volume of palisade and epidermal cells of the developing leaf of *Xanthium strumarium*. (From Maksymowych, R., 1963, *Amer. J. Bot.*, **50**, 891–901.)

work on leaf growth in *Xanthium* and Fig. 3–5 shows his data for cell expansion in palisade mesophyll and upper epidermis in that species. Cell enlargement is greatest in the epidermis and begins earlier than in the palisade, probably reflecting earlier cessation of cell division. But there is another difference; cell expansion in the epidermis is mainly in the horizontal plane whereas in the palisade it is mainly in the vertical. These changes begin at about LPI 0 so that cellular differentiation, in morphological terms, starts with the commencement of blade expansion.

The mechanisms which determine the differences in plane and timing of cell expansion in the layers of the leaf are not understood. When the primary leaves of *Phaseolus vulgaris* are kept in complete darkness, cells show the pattern of expansion characteristic of tissue type, although on a greatly reduced scale. This suggests that light is not the primary stimulus in this morphogenic response. Nor is the plane of division related to the plane of expansion for it is predominantly anticlinal in both epidermal and palisade layers. However, once it is established, the lateral expansion of the epidermis has important consequences for it is believed to be instrumental in separating the mesophyll cells which expand to a much smaller extent in the lateral plane and are thus pulled apart. This separation of cells is most clearly seen in the spongy mesophyll with its large air spaces, but it occurs to a significant extent also in the palisade mesophyll. Paradermal sections through the palisade cut parallel with the plane of the lamina, commonly show air spaces in this layer to account for 10–20% of the volume.

Table 3 The change in amount of major macromolecules during expansion of primary leaves of the french bean, *Phaseolus vulgaris*.

	Amount pg/cell	
Class of macromolecule	at unfolding	at full expansion
Nucleic acids (DNA and RNA)	70	220
Proteins	220	1320
Cell wall materials (including lignin)	250	2250

The rapid growth in size of leaf cells during blade expansion is associated with massive synthesis of macromolecules such as nucleic acids, proteins and cell wall materials (Table 3). Increase in the latter is especially marked, and continues after leaf expansion is completed as the wall continues to thicken and perhaps to extend vertically. Much of the material necessary for these syntheses is imported, and this gives special significance to differentiation of the vascular network of the leaf and in particular to the rapid development of phloem in the expanding leaf, as well as in the primordium.

3.5 Growth of the petiole

In non-sessile leaves, the petiole is usually visible in the older primordia prior to the commencement of unfolding. During expansion of the blade the petiole grows as well, increasing in length and circumference; these increases often continue after lamina expansion has ceased. They are important in establishing the leaf mosaic and positioning successive leaves so as to shade only minimally the older leaves lower down the stem (see Fig. 1–6).

Both cell number and cell size increase during petiole growth, and cell division may continue in the petiole for some time after it has stopped in the lamina. In the grass leaf, meristematic activity in the sheath intercalary meristem also continues for a short while after the blade meristem has ceased to function.

3.6 Growth regulators and leaf expansion

It is widely held that growth-regulating substances are involved in the control of development of a number of growing plant organs. In the case of leaves, evidence for this view is not very soundly based. It is true that young expanding leaves of many species often show high concentrations of gibberellin-like substances, which tend to fall during expansion, and that in many, but not all, young leaves levels of auxin (IAA) are also high. But such findings do not necessarily indicate that these substances play a specific role in maintaining growth of the leaf as a leaf; they could equally well be involved in non-specific ways characteristic of other growing systems, roots or buds, as well as leaves. What is certain though is that excised expanding leaf tissue requires presence of growth substances in the medium if growth is to proceed at rates which are comparable with those for intact leaves (Table 4). As might be expected high growth rates also require a carbon source and mineral salts to be provided.

Removal of the root system often leads to a reduction in leaf expansion. Since the roots are known to be sites of synthesis of cytokinins, it has been suggested that these compounds may be involved in leaf growth. Certainly, for some excised leaf material cytokinins promote expansion. However, it is difficult to

Table 4   The percentage increase in area of discs cut from young expanding leaves of the french bean (*Phaseolus vulgaris*) and cultured in the light for 24 h on various media.

	No Sucrose	*2% sucrose*
No additions	8	170
Plus mineral salts	17	174
Plus 10^{-6}M IAA	18	230
Plus 10^{-6}M GA_3	15	210
Plus mineral salts IAA and GA_3	25	285

disentangle whether or not such effects are specific, and whether these compounds are essential for expansion of a leaf growing naturally.

When applied exogenously, growth-regulating substances may affect leaf growth in two main ways:

(1) by changing the final size achieved

(2) by bringing about a change in leaf shape.

Effects on leaf size tend to be small or absent; thus spraying expanding leaves of *Phaseolus*, potato or barley with 100 ppm gibberellic acid (GA_3) increases area by 0–15%. In general, effects with GA_3 are greatest with early and frequent applications of the compound; leaves which have ceased to expand are unresponsive to treatment and in some species GA_3 application may actually reduce leaf area. Spraying with IAA and related auxins, with the exception of 2,4-D, has little effect on leaf growth unless concentrations are extremely high, as in some herbicidal preparations, when malformation of leaves may result. 2,4-D is effective in causing abnormalities of leaf growth when applied at very low concentrations and especially when applied to developing primordia where gross distortions may arise (Fig. 3–6). During subsequent unfolding, divisions in the interveinal regions are often severely inhibited, resulting in a strap-shaped leaf with fused or closely-appressed veins.

Application of growth retardants such as CCC (Cycocel, 2–chloro–ethyl-trimethylammonium chloride), as well as causing decreased stem elongation may also cause smaller leaves. Cell number in such leaves may be slightly

Fig. 3–6 Scanning electron micrograph of the stem apex of *Phaseolus vulgaris* treated with 2,4-D. The apical dome is seen below distorted primordia which have fused to form a ring which ultimately will give rise to a single strap-shaped leaf. (Photograph by Dr Fatima Aleixo-Pereira.)

reduced but cell size is more affected, with an increase in leaf thickness because of increased cell expansion in the vertical plane. Increased thickness and smaller size may cause treated leaves to appear darker although total chlorophyll content is usually similar to that of untreated leaves.

Effect of growth-regulating substances on leaf shape are discussed in more detail in Chapter 5.

3.7 Variegation and chimeras

The gardener has long cherished plants with variegated leaves. Variegations are caused by a number of agencies broadly classified as of genetic or non-genetic origin. The latter include variegation caused by physiological mechanisms such as mineral deficiency, or by virus infection. But it is variegation of genetic origin that is particularly interesting here, especially that involving chimeras which have contributed to our understanding of leaf growth.

The classical chimera possessed a goat's body with the head of a lion and the tail of a dragon. Plant chimeras are more mundane and arise when a cell in a meristem loses the potential to carry out a particular function and transmits this trait to its progeny. The most easily recognizable chimeras are those in which the ability to synthesise chlorophyll has been lost from some of the tissues. In an apical meristem, cells in all layers of tunica and corpus usually have the potential for chlorophyll synthesis. Even the outer layer of the tunica from which the epidermis is derived is genetically 'green'; as evidence for this, stomatal guard cells possess chlorophyll-containing plastids, and in many species, contrary to popular belief, other epidermal cells often contain small plastids. Now consider the case where a stem apex is composed of two tunica layers, L1 and L2, and a corpus, conveniently termed L3 even though it is not a single layer (Fig. 3–7).

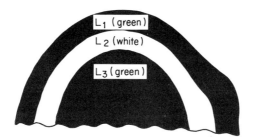

Fig. 3–7   Composition of the tissues of the stem apex of a green–white–green chimera.

Suppose a mutation to arise in L2 which prevents chlorophyll synthesis, then all tissues derived from that layer will be white and genetically the leaf will be Green (L1)—White(L2)—Green (L3). Leaves on such a chimera show a white margin of uneven breadth (Fig. 3–8a). If the chlorophyll deficiency arises in L3 (giving a genetic constitution Green—Green—White) the tissues of the leaf derived from that layer will be overlaid by green tissues so that the leaf will appear pale green with a dark green margin, derived from L2 (Fig. 3–8b).

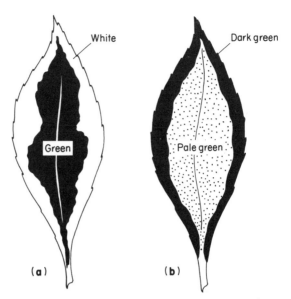

Fig. 3–8 Leaf appearance in a green-white-green chimera (**a**) and in one with the composition green-green-white (**b**).

The fact that such chimeras exist confirms a point made earlier (§2.2) that the layers of the leaf arise as a result of predominantly anticlinal divisions in the plate meristem. Occasionally, periclinal divisions do occur, and if these happen in L1 late in the period of cell division in a Green-White-Green chimera, a green speckling on the margin is sometimes seen; this is because small pockets of the L1 cells are found in the mesophyll layers. The frequency of periclinal divisions varies from zero to significantly high values; a frequency of periclinal: anticlinal divisions of around 1 in 3000 has been reported for a tobacco mutant.

Another important conclusion from studies on chimeras is that in the mature leaf, the contribution of cells from L2 and L3 can vary greatly in different parts of the leaf in an apparently random manner. This argues strongly against the idea that activity in the marginal and sub-marginal meristems controls lamina development.

The spectacular variegations in leaves of *Coleus* are not chimerical, but somatic. That is to say that there is apparently no heritable genetic difference between parts of the leaf of different colour. The variegation arises when during leaf development islands of cells produce certain pigment combinations. What determines the extent, stage and position of these islands is not known but the extreme variability of the patterning in such leaves, part of their horticultural attraction, suggests that the mechanisms involved are not closely linked to leaf development.

4 Environmental Effects on Leaf Growth

4.1 Light

The most spectacular effects of light on plant growth and development are seen when etiolated plants, grown in complete darkness, are compared with those grown in the light. The massive stem elongation which occurs in the dark is an obvious feature but failure of the leaves to expand is equally remarkable; they remain small and rudimentary, sometimes scale-like and of a different shape from 'normal' leaves on light-grown plants (Fig. 4–1). In such leaves both cell number and cell size are small, in contrast to the enormous cell expansion that occurs in etiolated stem tissue.

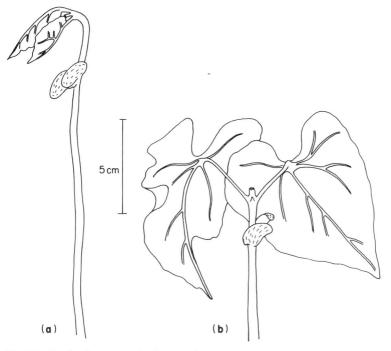

Fig. 4–1   Leaf and stem growth of plants of *Phaseolus vulgaris* grown in (**a**) darkness or (**b**) in full light conditions. (Drawings by Dr Diane Murray.)

The very different development of leaves in light- and dark-grown plants is due to the action of light in two major ways. The first of these involves the photomorphogenic pigment, *phytochrome* (KENDRICK and FRANKLAND, 1976), which can readily be shown to be implicated in many of the responses of dark-grown plants to low intensity red or white light. Phytochrome involvement in the growth of primary leaves of dark-grown *Phaseolus* plants is shown in Table 5. Exposure to a brief period of red light leads to a significant increase in cell number but this response is not found if the red light is followed by treatment with far red light, nor is it found with blue light. The effects on mean cell size are small and although treatment with red light leads to some leaf enlargement, values of leaf size, cell number and cell size remain very much less than in light-grown plants. To obtain full development of the leaves much higher light energies are required.

Table 5   Effects of 5′ treatments with light of different qualities on cell number and size in primary leaves of dark-grown plants of french bean, *Phaseolus vulgaris*.

Treatment	Cell number (millions)	Mean cell volume $(cm^3 \times 10^{-9})$
Dark control	15.6	4.2
Red	21.7	4.9
Red followed by far red	15.4	5.0
Blue	15.1	4.3
Values for plants grown in 12 h days of high light		
	44–50	75–90

Under most conditions of growth a developing seedling is subjected to low intensities of light some time before the plumule emerges above the soil level, since the soil surface tends to crack as the young plant grows. There is thus every likelihood that the low energy phytochrome responses are involved in controlling early development. Once emergence has occured the seedling will be exposed to higher light intensities, over long periods. The intensity, duration, and their product, quantity, of light, as well as quality are all important to lamina expansion.

Although different species vary widely in the responses of their leaves to light, work in controlled environments enables a generalized picture to be presented (Fig. 4–2). The main features are:

(1) an increase in area as intensity is increased, duration of the light period remaining constant

(2) a curvilinear or parabolic relationship between area and photoperiod, intensity remaining constant

(3) a more complex response of area to total quantity of light per day, depending upon the light intensity.

It must be stressed again that this interpretation is a general one and that considerable variation occurs between species.

Leaves grown in higher light intensities often have a larger cell number, cells being produced at a faster rate, and bigger cells than those grown in low light. So the larger area usually found comes about by increases in both the cellular

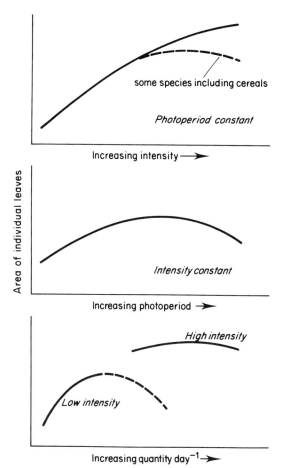

Fig. 4-2 The generalized relationships between leaf area and the light factors.

parameters. However, light intensity also has effects on the shape and thickness of leaves. In particular, high light conditions lead to a thick leaf with much larger cells in the palisade, while in leaves in low light intensity the epidermal cells tend to be greatly enlarged. Consequently caution is needed when attempting to relate leaf size to mean cell size in assessing light effects.

The rate of lamina expansion is often faster at higher light intensities (Fig. 4-3). This relects faster cell division and enlargement rates. It also reflects the fact that as the leaf unfolds it becomes progressively more self-sufficient for the metabolites, notably carbon assimilates, required for growth. In many species increasing the daylength results in an increase in leaf thickness. This is especially marked for succulents where it is often associated with a reduction in area.

In addition to direct effects upon leaf area, photoperiod may also exert

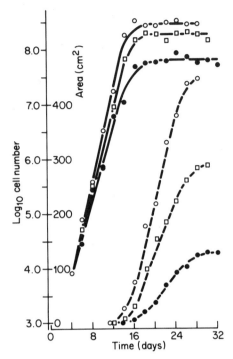

Fig. 4-3 Changes with time in area (– – –) and cell number of the second leaf of *Cucumis* grown in high (o), medium (□) and low light intensities (●). (From Milthorpe. F. L. and Newton, P., 1963, *J. exp. Bot.*, **14**, 483–95.)

indirect effects by affecting the onset of flowering (WHATLEY and WHATLEY, 1980). In many species, later-formed leaves are smaller in plants about to flower than in plants which remain vegetative; that is to say that there is an ontogenetic drift towards smaller leaves as flowering occurs. In photoperiodically-sensitive plants changing the daylength to induce flowering will promote this ontogenetic drift.

There are interesting effects of light quality on leaf growth which are of ecological significance. For a number of species characteristic of open habitats, increasing the ratio of far red to red light results in greater extension of stems and petioles and a reduction in area of individual leaves. Now, when daylight passes through leaves there is an enhancement of the far red:red ratio because of absorption of red light by the photosynthetic pigments. The morphological changes observed in plants grown in environments with a high far red:red ratio may therefore indicate a role for phytochrome in the detection of mutual shading between leaves and the initiation of responses to minimize this (HOLMES and SMITH, 1977).

For the whole plant, responses of leaf area to light are often complicated to interpret. This is because the rates of primordial production and of leaf unfolding are usually dependent on light intensity, and because intensity, and sometimes photoperiod as well, can affect the extent and pattern of branching and hence the number of stems on which leaves are borne. In general, an increase in light quantity, whether based on intensity or duration, will result in an increase in total leaf area of the plant.

4.2   Temperature

The response of most growing systems to increases in temperature in the range 10–35°C is to show an increase in growth rate, as the temperature rises to an optimum, followed by a decline (SUTCLIFFE, 1977). Leaf growth is no exception to this and an optimum temperature for lamina expansion can be found. This optimum, which is not always particularly sharp, is not necessarily the same for increase in total leaf area and for that of an individual leaf, since temperature may also have marked effects on the initiation of primordia and on the numbers of leaves unfolding.

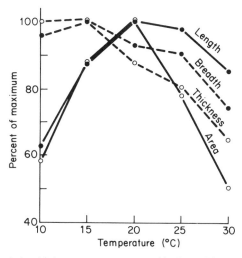

Fig. 4-4   The relationship between temperature and leaf growth in wheat. (Redrawn from Friend, D. J. C. Helson, V. A. and Fisher, J. E., 1962, *Can. J. Bot.*, **40**, 1299–311.)

Given the shape of the temperature response curve (e.g. Fig. 4–4) the *increase* in relative growth rate of the lamina with temperature with inevitably become smaller as the optimum temperature is approached. This is another way of saying that the temperature coefficient, Q_{10}, (SUTCLIFFE, 1977) will fall. These points are well shown by data for extension in length of poplar leaves (Table 6). There is an increase in **R** with temperature but a decrease in the size of the response to a 4°C rise between 16–20° and 20–24°C.

For an individual leaf, final size depends not only on rate of expansion of the lamina but also on the duration of the phase of enlargement. Consequently the optimum temperature for lamina expansion may not be the same as that for

Table 6   Relative growth rates for increase in length of *Populus* leaves at three temperatures, and the temperature coefficients for R. (Data from Pieters, G. A., 1974, *Meded. Landbh. Wageningen*, 74–106.)

Temperature (°C)	Relative growth rate day^{-1}	Q_{10}
16	0.116 ⎫	2.24
20	0.160 ⎬	
24	0.190 ⎭	1.38

final area. In primary leaves of *Phaseolus*, optimum temperature for leaf expansion rate is around 30°C, but the duration of expansion is reduced at this temperature so that the optimum temperature for final size is closer to 25°C.

For many leaves, shape is such that components which contribute to leaf size cannot easily be measured. For the grass leaf though, length, maximum breadth and thickness can readily be obtained, and the effects of temperature on these components determined. Date for wheat (Fig. 4–4) show a number of interesting features. For instance, although optimum temperature for leaf area is close to 20°, with a marked reduction at higher temperatures, length is much less sensitive to higher temperatures. Leaf breadth and thickness both show lower temperature optima, at about 15°C, with a steady decline in both parameters as temperature rises further. In this study, cell number per leaf varied only slightly and the effects were brought about mainly by changes in cell dimensions. Temperature thus affects the direction of cell growth and hence the shape of these leaves.

So far, in dealing with both light and temperature each factor has been considered in isolation from the other. This pragmatic approach masks the fact that light and temperature may interact in controlling leaf growth. A good example of this is shown by perennial rye grass, *Lolium perenne* (Table 7). At high temperature, 30°C, leaf area is reduced but the largest value is seen at the intermediate light intensity, whereas at 20°C the largest area was found with the lowest light intensity. Higher dry weight was found in the intermediate intensity

Table 7   Growth in area and dry weight, and the specific leaf area, for the fourth leaf of plants of *Lolium perenne* grown at two temperatures and three light intensities. (Data of Silsbury, J. H. 1970, *Trop. Grassl.*, **4**, 17–36.)

Temperature (°C)	*Area* (cm²) 20	30	*Dry weight* (mg) 20	30	*Specific leaf area* (cm² g⁻¹) 20	30
Irradiance (Wm⁻²)						
22	24.7	12.9	55.4	28.2	446	457
38	21.7	13.4	65.8	38.5	330	348
117	15.0	8.2	73.3	33.9	205	242

at 30°C, but in the highest intensity at 20°C. The ratio of leaf area to dry weight, *specific leaf area*, (HUNT, 1978) is unaffected by temperature at the low light intensities where the ratio is high (i.e. leaves are thin), but at the highest intensity the ratio is greater at 30°C. Again, it is likely that these differences are due to effects on cell size rather than on cell number.

The interactions between light and temperature make leaf growth studies in the field, where both factors are never constant, difficult to interpret. In consequence many workers prefer to use controlled environment facilities for experiments on leaf growth to ensure constancy of temperature and of light conditions so that the interactions between them can be more easily assessed.

4.3   Water

Growth of cells of the leaf, as of that of all plant cells, involves increase in water content brought about by uptake in response to a gradient of water potential between the cells and the water source, ultimately the soil. Water potential of an expanding leaf cell has two major components the osmotic, or solute, potential, and the pressure potential (wall pressure) (SUTCLIFFE, 1979). The solute potential will be governed by the flux of solutes into the cell and the rate at which they are converted into large polymers such as protein and cell wall materials. These last are particularly important because it is growth of the cell wall that allows a relief of the outwardly-directed turgor pressure and permits water uptake to occur. The changes involved can be represented diagramatically.

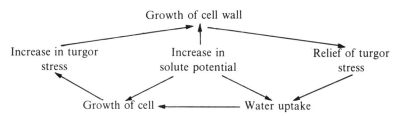

In fact, changes in turgor are hard to demonstrate in growing cells and the system must be one in which there are very rapid readjustments to changes in water content or turgor.

Leaf expansion is highly dependent on water uptake. ACEVEDO, HSIAO and HENDERSON (1971) found the rate of elongation of maize leaves to be almost stopped if the water potential of the solution around the roots was lowered by as little as 2 bars (= 200 kPa). This effect was rapid, but could be reversed if the water potential was raised to the original value (Fig. 4–5). Other evidence suggests that small reductions in relative water content of leaves (the amount of water held relative to that held at full turgor) do not affect expansion; however, larger reductions, to values less than 90%, quickly reduce cell enlargement which stops completely at values around 70%, due to reduction in turgor to a level where it is inadequate to maintain stretching and growth of the cell walls. Cell

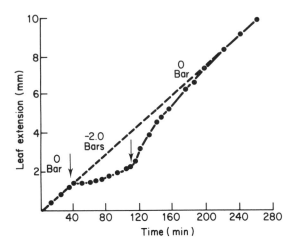

Fig. 4–5 The effect of lowering the water potential of the solution around the roots by 2 bars on elongation of a growing maize leaf. (Redrawn from Acevedo, E., Hsiao, T. C. and Henderson, D. W., 1971, *Plant Physiol.*, **48**, 631–6.)

division in the expanding leaf appears to be much less sensitive to water stress than cell expansion.

If plants are grown for prolonged periods under drought or water stress conditions, reductions in area of individual leaves occur and can be substantial (Table 8). Interpretation of these results is complicated by the fact that as well as direct effects of water stress on cell division and cell expansion there are also indirect effects on photosynthesis and translocation which will depress the amount of metabolites available for growth of the expanding leaves. Inhibition of the protein synthesizing machinery in the leaf cells, and reduced availability of mineral nutrients supplied through the roots are other effects of water stress which may affect leaf expansion.

Mention must be made here of adaptations of leaves of plants growing in xerophytic environments where water supply is limited. Characteristically such

Table 8 The effect of withholding water supply on leaf area in tobacco. (Data of Clough, B. F. and Milthorpe, F. L., 1975, *Aust. J. Plant Physiol.*, **2**, 291–330.)

Leaf number	Leaf area (cm²)	
	Watered controls	Unwatered plants
8	365	221
9	363	169
10	358	98
11	340	39
12	312	16

leaves tend to be thick and leathery with a well-developed cuticle. Lignified tissue is abundant in the form of sclerenchyma often associated with the vascular bundles which are prominent; even the epidermal cells may be lignified. Because of the absence of data from developmental studies on such leaves it is not known at what stage they become irreversibly committed to showing xeromorphic characters; there is evidence that when grown under conditions with abundant water leaves show greatly reduced xeromorphic features.

4.4 Mineral nutrition

The expanding leaf is a major sink for mineral nutrients, of which nitrogen, phosphorus, potassium and magnesium are especially important. The elements required for growth are supplied either directly from the soil solution via the xylem, or from older mature leaves by recycling. Import of mineral nutrients by the developing leaf can occur over prolonged periods. For example, in barley, leaves continue to be supplied with nitrogen, as nitrate, after full size is reached but export of reduced nitrogen, in the form of amino acids occurs concurrently so that there is no net increase in nitrogen. Leaves of cucumber (*Cucumis*) accumulate phosphorus until they are fully expanded but then become net exporters of this element even though significant amounts continue to be imported via the xylem.

Commonly, under poor conditions of mineral nutrition, fewer leaves are initiated and the areas of those that do expand are reduced (Fig. 4–6). In such

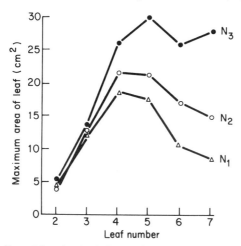

Fig. 4–6 The effect of three levels of nitrogen fertilizer on area of successive leaves of field-grown wheat. (Data of Puckridge, D. W. reproduced from Milthorpe, F. L. and Ivins, J. D., 1966, *The Growth of Cereals and Grasses*. Butterworths.)

cases there are reductions in both cell size and cell number in the leaf (Table 9) but it is not known whether the duration of leaf expansion is affected or whether nutrient supply directly affects cell division rates in growing leaves. Under more

Table 9   The effect of mineral nutrient supply on leaf area and cell number in *Cucumis*. Plants grown in large pots received about two and a half times as much mineral nutrient as those grown in small ones. (Data of Milthorpe, F. L. and Newton, P., 1963, *J. exp. Bot.*, **14**, 483–95.)

Pot size	*Leaf area* (cm²) small	large	*Cell number* (millions) small	large
4th leaf	171	484	328	498
5th leaf	136	437	282	454

extreme conditions of nutrient stress, chloroplast number in the mesophyll cells is reduced and this, together with reductions in chlorophyll content and changes in plastid fine structure, leads to a reduction in photosynthetic rate. This in turn may affect growth of younger leaves since the availability of assimilates required for their development may be reduced.

Effects of mineral deficiencies often show up quickly and characteristically in both growing and mature leaves (WALLACE, 1961; SUTCLIFFE and BAKER, 1981). For a number of crops analysis of leaf mineral content has been used to diagnose fertility and give an index of nutrient availability of the soil. Using such information the likely response to fertilizer applicant can be gauged.

4.5   Pests and diseases

When the growing plant is subjected to virus, bacterial or fungal attack, or to infestation by insect or other animal pests, leaf growth is often affected, and the number, size and even shape of leaves may be altered. Plant pathologists and entomologists are seldom interested in obtaining detailed growth measurements on leaves on the affected host plants and it is not always clear whether effects on leaves are directly ascribable to the pathogen or pest, or whether they are indirect and the result of damage to other parts of the plant. One case where the position is clear cut is that of yellow rust of wheat, caused by the fungus *Puccinia striiformis*. Here leaf growth of the host is affected and infection causes a reduction in leaf length and breadth as well as a reduction in leaf duration, since infected leaves senesce earlier. The fungus itself, only infects leaves which have emerged from the subtending leaves and are thus fully expanded so that the effect on leaf size must be indirect. Since the photosynthetic rates of infected leaves are significantly reduced, it is likely that the amounts of assimilate available for younger, developing leaves are also lowered bringing about the observed size reduction.

A reduction in the photosynthetic leaf area also results from the activity of leaf-eating animals of which the caterpillars of various species are the most important. But a large number of insect pests also feed directly on the stem apex of their host plants. Such a pest is the bug, *Lygus vosseleri* which feeds on cotton (*Gossypium*). By inserting its stylet into young unexpanded leaf primordia the insect causes considerable damage so that the leaf on unfolding presents a tattered appearance with a reduction in effective area of up to 60%. Associated with this damage is extensive abscission of developing flower buds, partly at least in response to the reduction in the assimilates available from the damaged leaves. The reduction in yield resulting from this shedding of buds can vary from 10 to 70%.

5 Leaf Shape

5.1 The establishment of leaf form

The enormous variety in leaf shape found in plants is very striking. Although the greatest variation exists between species, differences between plants of the same species are not difficult to find, and there may even be substantial differences in shape of leaves on the same plant. There is good evidence that leaf shape is controlled by genetic factors, often involving many genes, and that the effects of these can be modified by interaction with environmental factors such as light intensity, daylength or temperature.

Two lines of evidence indicate that major features of leaf form, e.g. whether it is compound or simple, entire or with deep lobing, are established very early in primordial growth. The first comes from direct observation. For example, the initiation of lateral leaflet and stipule primordia in *Phaseolus* occurs within 48 h of the first visible appearance of the original primordial structure (see Fig. 2–1). Another example comes from *Tropaeolum peregrinum*. The mature leaf of this species is deeply divided, with each lobe being further divided into lobules. Appearance of the major lobes begins when the primordium is about 200 μm long and development of the lobules commences when length is about 900 μm.

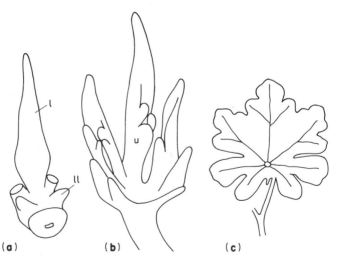

Fig. 5-1 Development of form in leaves of *Tropaeolum peregrinum*. The primordium (a) is 900 μm long and as well as the central lobe (l) small lobules (ll) are forming at the base. When the leaf is 2000 μm long (b) lobes and lobules are fully blocked out. Shape of the mature leaf is shown in (c). (Redrawn from Fuchs, C., 1975, *Ann. Sci. Nat. Bot. 12*ᶜ series **16**, 321–90.)

By the time that the central lobe measures 2000 μm, around 1/15 of its final size, the main features of shape are already blocked out (Fig. 5-1). Changes in the direction of growth, important in the initiation of the primordium itself, are also important in the development of lobes and lobules. The mechanisms involved in this directional control are not understood and this remains an important but difficult area for further work.

The second line of evidence showing leaf form to be determined very early, comes from surgical experiments. A classic series of such studies was performed by SUSSEX (1955) on the apex of potato (*Solanum tuberosum*). He made a series of incisions into the apex to isolate, in various ways, the site at which the next primordium would arise (I_1); the treatments are shown diagramatically in Fig. 5-2. Any cut which isolated I_1 from the apex led to initiation of a primordium

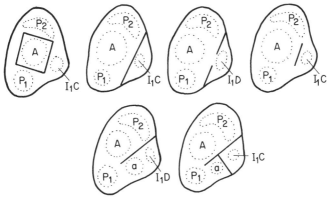

Fig. 5-2   The pattern of cuts used to isolate the apex of *Solanum tuberosum*. P_1 and P_2 are the two primordia already initiated when the cuts were made to isolate the aparal dome (A); I_1 is the incipient primordium which developed either as a dorsiventrally flattened leaf (D) or as a centric leaf (C); in some cases a second apiral dome is formed (a). (Redrawn from Sussex, I. M., 1955, *Phytomorphology*, **5**, 286–300.)

which gave rise to a radial, centric leaf. However, if the incision was incomplete, with a gap opposite I_1, or if it was made so obliquely that a new, lateral, apex was formed adjacent to I_1, then a normal dorsiventrally-flattened leaf was formed. Two main conclusions are possible from this work, firstly that the form of leaf, whether it is radial or dorsiventral, can be determined by events *before* primordial inception, and secondly that the apex itself plays an important part in specifying leaf form.

Surgical experiments of the sort performed by Sussex are never easy since even 'large' apices are still impossible to cut without the use of very small instruments and the dissecting microscope. The problem is even more acute when it comes to attempting surgery on the primordia themselves. However, SACHS (1969) was able to obtain interesting data by such dissections on primordia of pea (*Pisum sativum*). He removed various parts of leaf primordia when these were 70 or 30 μm in length, or younger still, when the next oldest primordium was 70 μm in length. He found that all except the most extreme treatments on the very

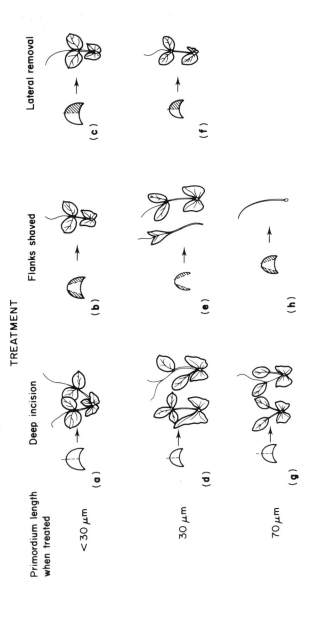

Fig. 5–3 The effect of surgical treatment of primordia of *Pisum sativum* at various developmental stages on the final form of the leaf produced. The leaves are normally of the form shown in (b) and (c). (Redrawn from Sachs, T., 1969, *Israel J. Bot.*, **18**, 21–30.)

youngest primordia led to development of a normal leaf, or of two normal leaves from the same point. When the primordia were 70 μm long surgery led to much more variation in the form of the 'leaves' produced (Fig. 5–3). These results can be interpreted as indicating that the older primordium is already structurally committed so that removal of a portion will remove a potential leaflet, or tendril, or stipule, which will then be lacking from the final structure. The youngest primordium, in contrast, is not determined in the same way and surgical treatments have no permanent effect because they are made too early and before the putative structures are established. Even so, it is clear from these results that leaf form is determined very early in primordial development.

5.2 Heterophylly and heteroblasty

A number of species found in aquatic and semi-aquatic habitats produce more than one type of leaf; they show *heterophylly*. Such plants include the water starwort (*Callitriche intermedia*) and several species of water buttercup, including *Ranunculus aquatilis, R. peltatus* and *R. pseudofluitans*. In these plants the form of submerged and floating leaves is very different, the former being narrower and often highly dissected, the latter broad and entire (Fig. 5–4). Such differences in shape are important adaptations to the environment, on the one hand reducing the resistance offered by the leaf to flow of water, and on the other presenting a large surface, capable of floating without becoming waterlogged.

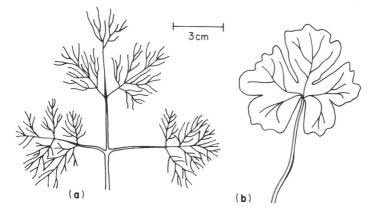

Fig. 5–4   Heterophylly in *Ranunculus aquatilis*; the dissected submerged leaf (**a**) and the floating entire leaf (**b**).

In several of the aquatic heterophyllous species, it has been shown by experiment that the type of leaf produced depends upon the environment to which the crown of the plant is exposed; moving the crown from the submerged to the aerial environment leads to the production of floating leaves, and *vice versa*. The response may also vary with photoperiod. Development in these species is clearly plastic and can be easily modified.

In other plants there may be gradual changes in the size and shape of successive leaves as the plant goes from the juvenile to the adult condition. This phenomenon is known as *heteroblastic development*. It is quite common among both annual and perennial herbs. Thus in the scots bluebell (*Campanula rotundifolia*), the first formed leaves are rounded and orbiculate, adapted to conditions of low light intensity amidst grassy competitors, but on the flowering stem the lower leaves are ovate and the upper ones narrow and linear. Here the heteroblastic developmental sequence is associated with the transition from the vegetative to the flowering state. This relationship between leaf shape and the onset of reproductive activity is seen in a number of species including morning glory (*Ipomoea cerulea*) and hemp (*Cannabis sativa*). In hemp (Fig. 5–5) successive leaves are more lobed and the margins become more serrated at succeeding positions. However, once flower initiation has commenced new leaves become smaller, less serrated and less lobed. Any treatment, such as exposure to short days, which accelerates flowering also accelerates transition through the heteroblastic sequence.

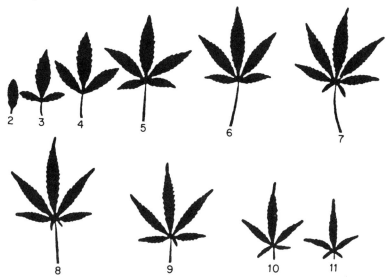

Fig. 5–5 The heteroblastic sequence in leaves of *Cannabis sativa* grown in 8 h days. (From Heslop-Harrison, J. and Heslop-Harrison, Y., 1958, *Proc. Roy. Irish Acad.*, **59** B, 257–83.)

Another species that shows substantial differences in the leaves of juvenile, non-flowering, and adult, flowering, plants is ivy (*Hedera helix*). The common climbing form is the juvenile phase with pubescent climbing stems, bearing many adventitious roots and producing 3–5 lobed palmate leaves with alternate phyllotaxis. Adult plants of ivy are shrubby and have smooth, upright stems, without adventitious roots, producing entire, ovate leaves with a spiral phyllotaxis. Cuttings struck from both forms give rise to juvenile plants and when adult stems are grafted on to juvenile stocks these too develop the juvenile habit. Application of gibberellic acid to adult plants causes the production of juvenile leaves and recent work has shown there to be high levels of gibberellin in roots of juvenile forms. It has been suggested that the transition from the juvenile to the adult form, which is unpredictable and may only occur after long period of time, depends upon a reduction in concentration of endogenous gibberellins to below a critical threshold level.

Application of gibberellic acid also promotes the juvenile leaf shape in a number of species including *Ipomoea*, but in others including *Eucalyptus* and the water fern *Marsilea*, the adult leaf form is induced by treatment. It is therefore clear that gibberellic acid cannot be considered as a juvenile hormone.

In conditions which promote rapid extension of the stem and shoot apex, such as high levels of mineral nutrition, low light intensity, and high night temperature, the juvenile condition is favoured. CUTTER (1965) has suggested that heteroblasty may involve a direct effect of the apex in the primordium which is dependent upon time and proximity. Where rapid stem and apical extension occurs primordia spend only a short time close to the apex and develop as juvenile leaves. Under less favourable conditions where growth is less rapid primordia spend more time adjacent to the apex and as a result grow as adult leaves. As a plant ages the size of the apex tends to increase and if its growth rate does not change then successive primordia must inevitably spend longer in its proximity and a transition to the adult condition would ensue. The nature of the interactions between apex and primordia which are required by this hypothesis are unknown; it is possible to invoke specific morphogens of the sort required by the field theory (§1.2) to explain how leaf form is determined but so far there are few direct clues to their existence, let alone their nature.

5.3 Sun and shade leaves

In considering leaf shape, thickness is often neglected. But it is important in determining the photosynthetic characteristics of the leaf, and often continues to increase after dorsiventral expansion of the lamina has ceased. Thickness can also be affected substantially by environmental factors, especially light. In many species leaves grown in high light intensities are 2–3 times thicker than those

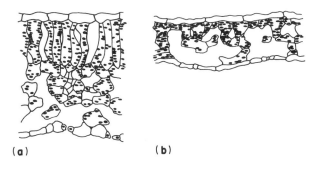

(a)   (b)

Fig. 5–6 Transverse section of lamina of *Impatiens parviflora* grown in full daylight (a) and in 70% daylight (b). (From Hughes, A. P., 1959, *J. Linn. Soc. Lond.*, **56**, 161–5.)

grown in low light and the effect of photoperiod on succulence has already been mentioned (§4.1). The increased thickness in the so-called sun leaves is due mainly to much greater development of the palisade mesophyll (Fig. 5–6) with concomitant increase in vein size and extent. Related to this are differences in the amount of air spaces within the leaf and in the ratio of the area of the internal surface of the mesophyll to the lamina area, which is much higher in sun leaves. One effect of this is to alter the resistances within the leaf to movement of CO_2 so that while both types of leaves show similar photosynthetic rates at low light intensities, shade leaves are not adapted to high light and have much lower photosynthetic maxima under such conditions (Fig. 5–7).

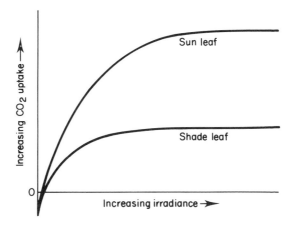

Fig. 5–7   The generalized relationship between CO_2 uptake and irradiance for sun and shade leaves.

Extreme forms of sun and shade leaves and intermediates between them can be found in large trees and shrubs when the light conditions vary within the crown. They occur also in herbaceous plants and can be induced by growing plants in high or low light intensities. Thus for sunflower (*Helianthus annuus*) a three-fold increase in light intensity at 24°C can reduce the specific leaf area from 600 to 450 cm^2 g^{-1} of leaf, significantly increasing leaf thickness.

The ontogeny of sun and shade leaves varies between species and in at least some trees the fate of a primordium to develop as a sun or shade leaf is determined within the bud. However, in many herbaceous plants it is the conditions obtaining during expansion which determine the type of leaf that develops.

6 The Functioning Leaf

6.1 The mature structure

Seen in the simplest terms, the leaf is an organ designed to achieve two major functions, the fixation of carbon dioxide in photosynthesis, and the export of the photosynthetic products. Leaf structure is closely adapted to these requirements. The more obvious adaptations include the dorsiventrally-flattened lamina of the typical mesomorphic leaf which presents a large surface for the interception of light, leaf orientation and display which reduces mutual shading between leaves, and the colour characteristics of the leaf ensuring absorption of much of the light of wavelengths used in photosynthesis. But there are other features important to effective leaf function. The high frequency of stomata in the lower epidermis, and often in the upper too, facilitates inward diffusion of carbon dioxide, and the inevitable outward movement of water vapour that occurs in transpiration (SUTCLIFFE, 1979). Related to this, the abundant air spaces within the leaf, accounting for 25% or more of the volume, allow ready diffusion of CO_2 and O_2 to all parts of the mesophyll. Because of its differentiation into palisade and spongy tissues the mesophyll itself is further adapted for gas exchange, presenting an internal cell surface area ten times or more larger than that of the lamina. For the export of assimilates, the vein network of the leaf ramifies in such a way that any mesophyll cell is close to the sieve element-companion cell complex of a minor vein, where loading of photosynthetic products into the phloem occurs for translocation.

6.2 The development of photosynthetic function

As it expands, the leaf shows an increasing rate of photosynthesis per unit area of surface, and since this too is increasing, the amount of carbon dioxide fixed per leaf rises rapidly. A maximum is reached at around the time of full lamina expansion or just before (Fig. 6–1).

The amount of CO_2 fixed in photosynthesis can be represented by a simple equation, analogous to that for Ohm's Law:

$$F = \frac{C}{R}$$

where F is the amount of CO_2 fixed per unit area and time, the photosynthetic flux density, in $kg\ m^{-2}\ s^{-1}$, C is the concentration gradient of CO_2 from the atmosphere external to the leaf, to the site of carboxylation in the chloroplast, in $kg\ m^{-3}$, and R, in $s\ m^{-1}$, is a composite term involving resistance to diffusion of

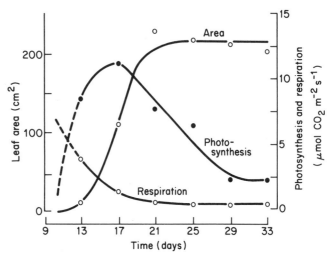

Fig. 6-1 Changes in area, photosynthesis and respiration rate in the expanding second leaf of *Cucumis sativus*. (Data of Hopkinson, J. M., 1964, *J. exp. Bot.*, **15**, 125–7.)

CO_2 into and within the leaf, as a gas, and in solution, and including also the capacity of the photosynthetic system to fix CO_2.

During early stages of lamina expansion R tends to fall as more stomata are differentiated and commence to function, and as enlargement and separation of the mesophyll cells increases the gas diffusion pathway and decreases the distance over which CO_2 moves in solution across and along cell walls and into the chloroplasts of photosynthesizing cells. At the same time there is an increase in the number and size of chloroplasts. In spinach, for example, chloroplast number per cell increases from 50 to 500 as leaf area increases from 1 to 50 cm²; in wheat, expanding cells of the mesophyll increase their chloroplast number by about four-fold to give a final average number of about 160 per cell. Coincident with these increases in number there are also increases in chloroplast size, and in the amount of chlorophyll and protein per chloroplast, as the organelle attains its fully-developed structure. A major result of these changes is the increase in capacity of individual chloroplasts to trap light and to carry out the carboxylation and other reactions involved in photosynthesis. Together, the increases in chloroplast numbers, size and function, also contribute to a decrease in R as the leaf expands.

At around the time that full leaf expansion is achieved R begins to increase again and photosynthetic activity of the leaf falls as a result. Part of the reason for this is a gradual decrease in the amount of protein per leaf and per chloroplast, but changes in cell wall thickness which increase the pathlength for movement of CO_2 in solution may also contribute.

During the early stages of leaf growth dark respiration rates are high, characteristic of a rapidly growing meristematic tissue, but tend to fall as expansion proceeds and continue to fall in the mature leaf (Fig. 6–1). However in those plants which fix CO_2 into triose phosphate, so-called C_3 plants, the decline in dark respiration is more than offset by a rise in the rate of photorespiration. This is an oxygen-dependent breakdown of a proportion (up to 50% in some species) of the recent products of photosynthesis. As the name suggests this form of respiration occurs only in the light, and it does not occur in those plants, such as maize, in which CO_2 is fixed into C_4 compounds.

6.3   Translocation and leaf development

From primordial initiation until the time when lamina expansion is well advanced the growing leaf is heterotrophic and entirely dependent upon metabolites imported from older leaves. Gradually, as photosynthetic activity develops, the leaf becomes a net producer of fixed carbon and import becomes progressively less important, with export increasing steadily. Fig. 6–2 shows this

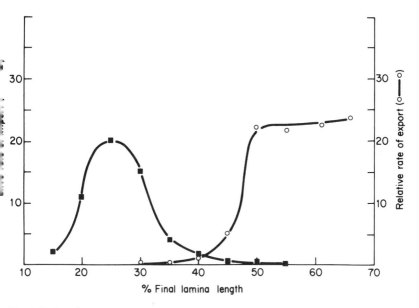

Fig. 6–2   Relative rates of import and export by the expanding 7th leaf of sugar beet. (Data of Fellows, R. J. and Geiger, D. R., 1974, *Plant Physiol.*, **54**, 877–85.)

sequence for the seventh leaf of sugar beet. For this leaf maximum rate of import occurs at LPI 0.5 when the leaf is about 25% of its final size. Soon after this, export of assimilate begins so that there is a phase when the leaf is simultaneously exporting and importing material. Since the tip of the leaf matures first import ceases there earlier than in the later maturing basal regions. Conversely photosynthetic products from the tip are the first to be exported from the leaf. Not that all the locally-produced assimilates will be exported;

some will be retained for local use in, for example, the synthesis of cell wall material, and in some species, probably the majority, assimilate from older parts of the leaf may be used in growth of the younger basal regions.

In sugar beet, the duration of the phase of simultaneous import and export by the developing leaf is quite long extending over nearly 2 LPI units, and import only ceases when the leaf is about 50–60% of its final size; it is at about this time that the rate of translocation out of the leaf is maximal. In other species, maximal rates of translocation may occur only when the leaf is fully expanded. Fig. 6–3 brings together data on the carbon balance for expanding leaves and shows the general interrelationships between import and export of carbon, photosynthesis and leaf growth.

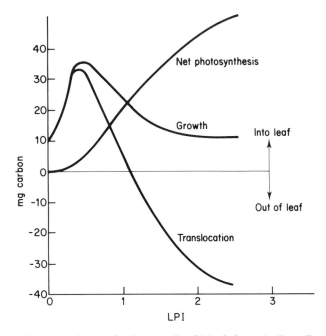

Fig. 6–3   The carbon balance for the expanding 5th leaf of squash. (From Turgeon, R. and Webb, J. A., 1975, in *Phloem Transport*, ed. S. Aronoff *et. al.* Plenum Press.)

The development of the vascular system is closely related to the pattern of translocation in the leaf. Extension of the procambial strand from the base of the primordium is rapid and acropetal development of the major veins is associated with import during early growth. The export of photosynthetic assimilate begins with loading into the minor veins of the fifth order or so, and these develop from the tip, basipetally. In a number of species, including sugar beet, the companion cells in the minor veins are much larger in relation to the sieve elements than in the major veins and this may be related to their involvement in the loading process. The extent of the minor vein network in the typical leaf is prodigious. In sugar beet (GEIGER, 1975) there is about 70 cm of minor vein for every cm² of leaf and the surface area of the sieve element companion cell complexes in these veins is about 88 mm² cm⁻² of leaf. This network extends over all parts of the leaf and it is estimated that there are no palisade cells more than about 100–120 μm from a minor vein, and that about 30 cells supply assimilate to a portion of minor vein 30 μm in length. How far the initiation of export is dependent upon structural maturation of the minor veins is not completely certain. It seems likely that physiological differentiation and development to provide the mechanism for loading sugars into the phloem may occur more slowly than visibly-obvious anatomical maturation.

6.4 Sensitivity to the environment: photoperiodism

Mature leaves are involved in a vast range of metabolic and physiological activities in addition to their photosynthetic and translocation functions. For example, in many species, leaves are major sites for the reduction of nitrate, and there is good evidence that leaves are sites of synthesis of plant growth substances including gibberellins; leaves are often involved in the production of secondary metabolites such as the essential oils which give a characteristic scent to many plants. The level and extent of these metabolic activities will be affected by environmental factors such as temperature and light quantity and quality.

Leaves also have an important role in flowering of the species which are photoperiodically-sensitive. For more than 50 years it has been known that the leaves are the sites of perception of the photoperiodic stimulus in such species and that if leaves are removed flowering will often be prevented. In the cocklebur (*Xanthium strumarium*) the classic plant of photoperiodism studies, half-expanded leaves are the most sensitive to daylength, but in many plants fully-expanded leaves are also effective. The amount of leaf that has to be exposed to inductive conditions is often quite small, thus in *Xanthium* 7 cm² in area is sufficient, and in *Lolium temulentum* as little as 3 cm² is enough. What happens in the leaf during the inductive conditions remains one of the great mysteries of plant physiology, but it is believed that either a new metabolic activity is initiated, or an existing one modified, to produce a factor or factors which bring about flowering. These factors, about whose nature we are ignorant, are produced in the leaf and have to be translocated out, through the phloem, to meristems where flowers are to be formed. Because of this involvement of phloem transport the apparent sensitivity of the leaf to photoperiod may depend upon the development of an export capacity.

6.5 Senescence

From the initiation of a leaf primordium to full lamina expansion can take from as little as a few days in small annual herbs to as long as four years in the case of leaves of the cinnamon fern (*Osmunda cinnamomea*) where the primordium is especially slow growing. Longevity of the expanded blade also varies greatly. The leaves of most herbaceous and deciduous angiosperms persist only for one season or a part of it, but most gymnosperms and many tropical broad-leafed trees, some palms and many ferns, have leaves which persist for several years. The needle leaves of the scots pine (*Pinus sylvestris*) are retained for three or sometimes four years before they absciss and are shed. The leaves of the monkey puzzle tree (*Araucaria araucana*) may persist for more than 8–10 years, and in some palms leaves may be retained for similar periods before eventually withering and disintegrating.

The longevity of leaves, as of their parent plants, is probably determined only indirectly by genetic factors although these do exert more direct control over the processes of *senescence*, which can be defined as the deteriorative changes in an organ or organism which lead ultimately to death. In the case of leaves these changes are well-known if poorly understood, and can be recognized as beginning as soon as, or even before, the leaf has reached full size. The decline in photosynthesis in the expanded leaf has already been mentioned. Leaf senescence is associated with a gradual fall in chlorophyll and protein (Fig. 6–4), especially Fraction I Protein, the key CO_2-fixing enzyme ribulose bisphosphate carboxylase. Gross content of ribose nucleic acids goes down too, although some types of RNA may actually increase during senescence. As might be expected, degradative enzymes such as protease and ribonuclease increase in amount or activity, and coincident with this, amounts of rough endoplasmic reticulum and numbers of ribosomes start to fall. Ultimately chloroplast membranes become disorganized, nuclei degenerate and the leaf dies.

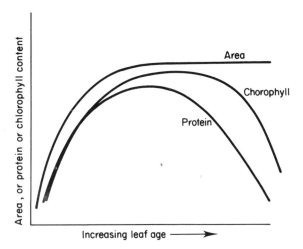

Fig. 6–4   Generalized diagram of changes in protein and chlorophyll content throughout the life of a leaf.

During senescence translocation out of the leaf continues and many of the products of the breakdown processes as well as mineral elements such as potassium, are transported to other growing organs including young leaves and developing fruit. So the process is not entirely wasteful. Nor is it inevitable and uncontrolled. It has long been known that leaf senescence can be delayed by removing fruit from a plant, and even decapitation and removal of the developing younger leaves can sometimes be effective in reducing senescence. Conversely, when a leaf is removed from the plant it usually shows enhanced senescence, often seen as a faster breakdown and loss of protein. These observations suggest that other parts of the plant may regulate changes in the leaf and plant growth-regulating substances have been implicated in this.

More than 20 years ago, cytokinins were found to delay senescence in detached leaves and leaf disks. Gibberellins and even auxin have also been shown to be effective in retarding senescence in certain species. The evidence suggests that growth substances may affect rates of synthesis of protein and RNA in the leaf, but they also reduce the rate of breakdown by preventing the rise in ribonuclease and protease activity and it is this role that may be the more important where detached leaf tissue has been studied. Unfortunately, these *in vitro* studies, suggestive though they are, give no real evidence of the control mechanisms operating in the intact plant. It is still uncertain for example, whether senescence is enhanced directly by substances coming from developing organs, or whether developing organs act as sinks accumulating anti-senescence

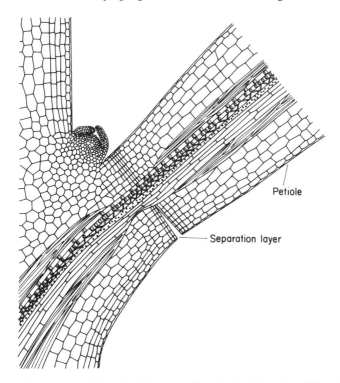

Petiole

Separation layer

Fig. 6–5 The structure of the abscission zone. (Drawing by Mrs A. B. Addicott.)

factors coming from elsewhere in the plant, say the roots, thus depriving the mature leaves of them.

6.6 Abscission

Senescence changes lead ultimately to death of the leaf. In many annual herbs leaves persist and wither in position and this is also the fate of the long-lived leaves in certain palms. In deciduous trees however, leaves are shed by regulated processes of abscission. The scale of abscission can be enormous and in Canadian deciduous forests around three tons of leaves are shed per hectare per annum in what is appropriately called 'the fall'.

Abscission occurs as a result of changes in the *abscission zone*. This is found at the base of the petiole in simple, entire dicotyledonous leaves, and additional abscission zones are found at the base of the leaflets in many compound leaves. The abscission zone (Fig. 6–5) is often recognizable because of a slight constriction at the leaf base and this region may also be of a different shade of colour from the surrounding tissues. Internally, the zone consists of rather small cells with thin walls and often lacking starch. Characteristically, lignified cells are absent, except for the xylem elements and occasionally a few fibres. The abscission zone is anatomically distinct from the adjacent petiolar tissues and is considered a region of structural weakness.

Prior to shedding of the leaf from the stem a separation layer develops in the distal region of the abscission zone; proximal to this, cells often become suberized and lignified. In this way a protective layer is formed which acts as a seal to the surface exposed when separation is complete. The process of

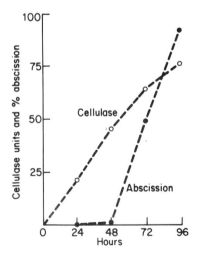

Fig. 6–6 The changes in cellulase activity and abscission in the abscission zone of explants of the petiole of orange. (From Ratner, A., Goren, R. and Monselise, S. P., 1969, *Plant Physiol.*, **44**, 1717–23.)

separation is brought about by the lysis and breakdown of middle lamella and primary cell walls of the parenchymatous cells in the separation layer. Increases in activity of the enzyme pectinase in the abscission zone region are associated with this (Fig. 6–6). Separation proceeds from the outside of the petiole inwards, and eventually only the lignified cells of the vascular bundle remain intact. These cells are broken by mechanical forces, often by wind, and the leaf falls.

All the major groups of plant growth substances have been shown to affect abscission when applied to suitable test systems, but IAA and ethylene are believed to be the most important compounds controlling natural abscission. Using explants of petiole and adjacent stem tissue, it has been shown that IAA is initially inhibitory to abscission, but later it will promote abscission. In both these stages, application of IAA stimulates formation of ethylene in the tissues. Ethylene itself strongly promotes abscission when applied in the second phase, but is less active in the first. It appears to cause an increase in the cell wall degrading enzymes in the separation zone. The sequence of events leading to abscission can be summarized as follows: the leaf is initially a source of IAA which is inhibitory to abscission; as it ages and senescence proceeds, IAA production declines and the production of ethylene, itself an inhibitor of auxin transport out of the leaf, increases. The leaf now enters a phase where, because of falling levels of IAA and increasing amounts of ethylene, cells destined to form the proximal protection zone become suberized and in those of the separation layer the activity of cell wall hydrolytic enzymes increases. Cell wall breakdown in the abscission zone follows and separation occurs.

Further Reading

General references

CUTTER, E. G. (1971). *Plant Anatomy, Part 2. Organs*. Edward Arnold, London.
DALE, J. E. and MILTHORPE, F. L. (eds) (1982). *The Growth and Functioning of Leaves*. Cambridge University Press, Cambridge.
ESAU, K. (1965). *Plant Anatomy*, 2nd Edition. Wiley, New York.
EVANS, L. T. (1975). *Daylength and the Flowering of Plants*. Benjamin, Meres Park.
MAKSYMOWYCH, R. (1973). *Analysis of Leaf Development*. Cambridge University Press, Cambridge.
MILTHORPE, F. L. and MOORBY, J. (1979). *An Introduction to Crop Physiology*. 2nd Edition. Cambridge Univerisity Press, Cambridge.
WILLIAMS, R.F. (1975). *The Shoot Apex and Leaf Growth*. Cambridge Univeristy Press Cambridge.

Studies in Biology Series (all published by Edward Arnold, London)

HUNT, R. (1978). *Plant Growth Analysis* no. 96.
KENDRICK, R. E. and FRANKLAND, B. (1976). *Phytochrome and Plant Growth* no. 68.
LANGER, R. (1979). *How Grasses Grow*, 2nd Edition no. 34.
SUTCLIFFE, J. F. (1977). *Plants and Temperature* no. 86.
SUTCLIFFE, J. F. (1979). *Plants and Water*, 2nd Edition no. 14.
SUTCLIFFE, J. F. and BAKER, D. A. (1981). *Plants and Mineral Salts*, 2nd Edition no. 48.
WHATLEY, J. M. and WHATLEY, F. R. (1980). *Light and Plant Life* no. 124.

Other references

ACEVEDO, E., MSIAO, T. C. and HENDERSON, D. W. (1971). *Plant Physiol.*, **48**, 631–6.
AVERY, G. S. (1933). *Amer. J. Bot.*, **20**, 565–92.
CUTTER, E. G. (1965). *Bot. Rev.*, **31**, 7–113.
DALE, J. E. (1976). *Cell Division in Leaves* in *Cell Division in Higher Plants* (M. M. Yeoman, ed.). Academic Press, London.
DOODSON, J. K., MANNERS, J. G. and MYERS, A. (1964). *Ann. Bot.*, **28**, 459–72.
ERICKSON, R. O. and MICHELINI, F. J. (1957). *Amer. J. Bot.*, **44**, 297–304.
FUCHS, C. (1975). *Ann. Sciences Nat., Bot.* 12ˢS., **16**, 321–90.
GEIGER, D. R. (1975). *Phloem Loading* in *Transport in Plants 1. Phloem Transport* (M. H. Zimmerman and J. A. Milburn eds). Encyclopedia of Plant Physiology N.S. Volume 1. Springer, Berlin and New York.
HELLENDOORN, P. H. and LINDENMAYER, A. (1974). *Acta Bot. Neerl.* **23**, 473–92.
HOLMES, M. G. and SMITH, H. (1977). *Photochem. Photobiol.*, **25**, 551–7.
KAPLAN, D. R. (1973). *Quart. Rev. Biol.*, **48**, 437–57.
KUEHNERT, C. C. (1972). *Determination of Leaf Primordia in* Osmunda cinnamomea in *The Dynamics of Meristem Cell Populations* (M. W. Miller and C. C. Kuehnert eds). Plenum New York.

LYNDON, R. F. (1976). *The Shoot Apex in Cell Division in Higher Plants* (M. M. Yeoman, ed.). Academic Press, London.

RICHARDS, F. J. (1951). *Phil. Trans. Roy. Soc. B.*, **235**, 509–64.

SACHS, T. (1969). *Israel J. Bot.*, **18**, 21–30.

SCHWABE, W. W. (1971). Chemical modificatons of phyllotaxis and its implications in control mechanisms of growth and differentiation. *Symp. Soc. exp. Biol.*, **25**, 301–22.

SUSSEX, I. M. (1955). *Phytomorphology*, **5**, 286–300.

THORNLEY, J. H. (1976). *Mathematical Models in Plant Physiology.* Academic Press, London.

VEEN, A. H. and LINDENMAYER, A. (1977). *Plant Physiol.*, **60**, 127–39.

WALLACE, T. (1961). *The Diagnosis of Mineral Deficiencies in Plants.* H.M.S.O., London.

Subject Index